AF864157

Creation By Evolution

High Sierra Books
Gold Beach, OR 97444

Published by:

BILL HUTH PUBLISHING
3500 75th St. West
Rosamond, CA 93560
1 (800) 758-0771
Fax (661) 256-9140
www.billhuth.com

ISBN: 0-9665379-9-8
LCCN: Pending

FORWARD

First published in the 1920s, *Creation By Evolution* is a compilation of papers written by prominent scientists and philosophers who made the first attempt to emphasize the benefits of the theory of evolution to a world which was (and is) locked in furious debate over the seemingly unbridgeable chasm between religion and science. The book was put together just after Darwin had completed his works on evolution and the *Origin of Species*. Feeling that the foundations of their beliefs were under mortal attack, many religious leaders and devout creationists railed against the blasphemous theories put forth by scientific inquiry.

The writings contained in this work not only provide a fascinating look at a world rendered by the religious/scientific debate which combusted after Darwin's unveiling of his theories, but also show the insight of thinkers who saw that the benefits inherent in the theory of evolution were not necessarily at odds with religious doctrine. *Creation By Evolution* is a historical document which, for obvious reasons, must be preserved and studied; it also chronicles the first attempts of prominent thinkers to provide a synthesis to the dialectic of science and religion. Moreover, the publisher's commentaries and observations interwoven throughout the text provide a bridge between the past and future.

Past and future. Science and religion. These antithetical concepts seem to be locked in permanent opposition. A lifelong advocate of evolution *and* creationism, Huth provides a unique and startling vision of the human position in the universe. Like Kant, Huth's belief that time and space are man-made constructs, the past and future illusion, science and religion interpretations of the same phenomena, provides validity to the notion that nothing is inseparable. For Huth, God (for lack of a better term) progresses with man through evolution. Scientific and all other inquiry into the nature of existence serve to further the purpose of that which is present in all of the physical emanations we have aptly coined *Universe*.

Spirit (again for lack of a better term) and matter. Another opposition? The suggestion that matter is the physical manifestation of spirit is not new. However, with the advent of the theory of evolution these early writers actually caught a glimpse of the God in the machine. Using empirical evidence and scientific method, these thinkers liberate their fellow man from the constraints of religious dogma and stagnant morality. If all matter is a manifestation of spirit, and man, made of matter, has evolved in such away that the brain, also made of matter, is able to conceive of the consciousness inherent in itself and in all of creation, then the entire process of evolution might be aimed at the realization that, through matter, the great unconsciousness that is spirit becomes aware of itself. The physical universe is the path to the God becoming self-aware. God needs self-validation.

The process of evolution is very long by human standards. In the end, however, because energy (spirit) can neither be created nor destroyed, and is therefore eternal, time is irrelevant. What is a trillion trillion years in the face of eternity? Energy is self-contained, unbound from the shackles of time and space.

By adopting the theory of evolution, the writers contained within this book have opened the door to the possibility of being. Humanity, freed from the superstitions of a gloriously inglorious past, can play a dynamic role in the process of transformation. No longer is the individual subjugated to a wrathful deity and mechanical predestination. "The Kingdom of Heaven lies within your heart," in this time, upon this earth. If only you are able to dedicate yourself to the search for the key which will open the door and awaken eternity.

Eternity sleeps deep. The waking of the Great Spirit into self-consciousness is no easy task. With the advent of scientific method and the theory of evolution, again however, we have strode giantlike in the direction of the God in man. Quantum physics gets ever closer to a legitimate understanding of the mysterious processes of physical being. The microcosmic universe contained within the atom may reveal macrocosmic answers to questions long considered to be the property of religious leaders.

It is obvious that concepts evolve just as physical organisms. "Over two thousand years and no new gods," writes Nietzsche. To create one must synthesize the old theses and antitheses. The old standard of science (as suggested by quantum theory) has reached its zenith. Old modes of religious dogma have outlived their usefulness. Only through psychological transformation, a new way of thinking, can these two stalwarts finally be laid to rest in order that a new Phoenix can rise. Religion and science will be one, and we will be one step closer to the realization that *all* is one.

We owe a great debt to explorers such as Darwin, Jesus, Kierkeggard, Goethe, Buddha, Aristotle, Rilke, Jung, and Nietzsche, who, in large part, have delivered us to this exciting point in the process. I would like to pay tribute to the writers contained in *Creation By Evolution*, and especially Bill Huth, the publisher of this historically important work. By adding his own insights into the book, Huth infuses the writing with a new vibrancy, a relevancy which is integral to our own and future generations.

This work is both a historical goldmine and a new release. Another chasm bridged. Huth's tireless quest for truth stirs the eternal unconscious into waking. It is up to you and I to take it from here.

O Man! Attend!
What does the deep midnight's voice contend?
"I slept my sleep,
And now awake at dreaming's end:
The world is deep,
Deeper than the day can comprehend.
Deep is its woe,
Joy – deeper than heart's agony
Woe says: 'Fade! Go!'
But all joy wants eternity,
Wants deep, deep, deep, eternity!"

Nietzsche, *Thus Spake Zarathustra*

John Gist, 2000

Major Contributors

(Dates correspond to original publication)

Bather, Francis Arthur-
Professor of Geology, University of Glasgow
Rolleston Prize of Oxford and Cambridge for research in Biology

Berry, Edward Wilbur -
Professor of Paleontology, John Hopkins University
Walker Prize, Boston Society of Natural History

Conklin, Edward Grant -
Professor of Biology, Princeton University

Gager, C. Stuart -
Director Brooklyn Botanical Garden
Member Botanical Society of America
Fellow American Association for the Advancement of Science

Gregory, John Walter -
Professor of Geology, University of Glasglow
Victoria Medal, Royal Geographical Society
Gold Medal, Royal Society, Edinburgh
Bigsby Medal, Geological Society, London

Gregory, William King -
Professor of Vertebrate Paleontology, Columbia University
Curator of the Department of Comparative Anatomy and Ichthyology, American Museum of Natural History

Holmes, Samuel Jackson -
Professor of Zoology, University of California
Fellow American Association for the Advancement of Science

***Huth, Bill** - (as of 2000)
Owner, Willow Springs International Raceway
Publisher, High Sierra Books

Huxley, Julian Sorell -
Professor Zoology, Kings College, London
Newdigate Prizeman in Natural Science

Jennings, Herbert Spencer -
Professor of Zoology, John Hopkins University
President American Zoological Society

Jordan, David Starr -
Chancellor Emeritus, Leland Stanford University
President American Association for the Advancement of Science 1909-1910

Loomis, Frederick Brewster -
Professor of Geology, Amherst College
Fellow Geological Society of America

Lull, Richard Swann -
Professor of Paleontology, Yale University
Director of Peabody Museum

MacBride, Ernest William -
Professor of Zoology, Imperial College, London
Vice President Zoological Society, London

Morgan, C. Lloyd -
Professor Emeritus, University of Bristol
Vice-Chancellor, University of Bristol

Newman, Horatio Hackett -
Professor of Zoology, University of Chicago
Fellow American Association for the Advancement of Science

Osborn, Henry Fairfield -
President of the American Museum of Natural History
Research Professor of Zoology, Columbia University

Parker, George Howard -
Professor of Zoology, Harvard University
Foreign Member Linnaean Society, London

Poulton, Edward Bagnall -
Hope Professor of Zoology, University of Oxford
President of the Entomological Society, London

Scott, William Berryman -
Blair Professor of Geology and Paleontology, Princeton University
Wollaston Medal, Geological Society, London

Sherrington, Sir Charles Scott -
President Royal Society
Royal Medallist, Royal Society
Professor of Physiology, Oxford University

Shipley, Sir Arthur Everett -
Vice-Chancellor, Cambridge University
Master of Christ's College, Cambridge

Smith, G. Elliot -
Professor of Anatomy, University of London
Royal Medal, Royal Society

Thomson, J. Arthur -
Regius Professor of Natural History, Aberdeen University
Terry Lecturer, Yale University

Watson, David Meredith -
Jordell Professor of Comparative Anatomy, University of London

Wheeler, William Morton -
Professor of Entomology, Harvard University
Dean of the Bussey Institution for Research in Applied Biology, Harvard University
Member American Philosophical Society

Woodward, Sir Arthur Smith -
Past President of the Linnaean an Geological Societies of London
Royal Medal, Royal Society
Lyell Medal and Wollaston Medal, Geological Society, London

YOU AINT NOTH'N BUT A HOUNDOG

Today in this very great 20th century, all mankind are being enlightened with the facts of evolution.

It has become a positive fact, that a few billion years ago the planet earth had a temperature on its surface of well over several thousand degrees.

After many billions of years have elapsed, this great sphere has cooled to a temperature, which has enabled moving creatures with a body of flesh to live and produce on its surface.

Through ignorance, the great speakers of a few thousand years ago could not conceive man as being in any form other than the one they were then seeing reflected by their early civilization.

Today this same ignorance is still confusing the multitudes. More arguments have had their beginning with "Adam," than can possibly be mentioned. Still there are becoming more each day, who are positive that God created "Adam" in the shape of man today. If they would read the bible with a clear mind, it is very simple to see the truth.

To persuade the population of our current civilization that, they are related to the apes is quite a task. These die-hards are still positive as they have been taught, that God created man as the leader of all life on earth and that he was created as he is, and that's it.

Not only have you evolved through the stages of prehistoric man, but you have CRAWLED ON ALL FOURS! Before this, you lived as THE TOAD, and other creatures who crept out of the great "FLOOD," which once covered our earth. And about this flood. What other excuse could the religious leaders make than "NOAH'S ARK," when they stumbled across sea shells high on the mountain tops. Through those who were not satisfied with this "excuse," men of science have finally exposed the truth.

Not only have you evolved through life from the TINY NEUTRON, but today you stand billions of years from your UNCONCEIVABLE FUTURE.

During the time of Columbus, the masses felt the EARTH TO BE FLAT. Today we not only understand the earth is round, but we have proof of that fact in pictures which were taken from rockets which have been fired hundreds of miles into the atmosphere.

C.C.H. VOICE OF CREATION
1957

ACKNOWLEDGEMENTS

This book was originally published in the mid 1920's. I feel that Frances Bacon in an attempt to expose mankind to the benefits of Evolution edited and compiled the information which it contains. Never was it his intention to destroy any of the worlds great selection of religions, but only give humankind a better base to build a future. His production of these contents came at a time when Charles Darwin had just completed his works on Evolution and The Origin Of Species, which had set the religious and scientific world on fire with heated debates and a raving voice from all directions.

The accompanying authors who endeavored to, in a simple form of dialogue, interpret Creation By Evolution, the original title of this works, can now be much more appreciated than the time of its first publication; their sounding board and our mother earth respond to human kind, who is searching for the truth. Together they make up a Voice Of Creation which shall not be silenced. Others will follow.

BILL HUTH

When will the lions lay down with the sheep?
After they've had a few!

TABLE OF CONTENTS

TABLE OF CONTENTS, continued

TABLE OF CONTENTS, continued

TABLE OF CONTENTS, continued

ABOUT THIS BOOK

THIS book is the result of a wish to obtain the judgment of leading scientific scholars of the English-speaking world concerning our present knowledge of how living things in Nature come about,—to obtain actual evidence of Nature's method of creation.

Each of the writers gives an independent record of research in his own particular field, striving to do so in language that all may understand and appreciate.

All these scientists of the organic world, though studying from many different points of view, find that creation is a gradual process of transformation, or evolution, each condition the outcome of things preceding, according to natural laws.

Nature shows nothing finished and perfect in the beginning; she shows orderly divergence and an advance from lower to higher levels of creation.

The book does not attempt to explain the origin of life, or to determine the causes that lie behind the changes in living things from age to age. It attempts to show that there are changes and to describe how they come about.

The revelation of creation by evolution which comes to us through science widens and exalts our outlook on life and our religious faith, and these papers have been assembled in the hope that they may lead to a more general understanding of Nature and Nature's Way.

To those who have set forth for the reader the hard-won results of their life's research I tender my sincere thanks and deep appreciation.

FRANCES MASON.

INTRODUCTION

"Since He that made us with such large discourse,
Looking before and after, gave us not
That capability and God-like reason
To rust in us unused."

Shakespeare.

"MAN looks before and after," and, peering into the darkness of the past, has often sought answer to the question how he came to be. He has felt that knowledge of the process which has underlain his making, of whence he came and by what route he has reached his present station, should set that station in a clearer light for his contemplation and should afford him, possibly, some glimpse of his terrestrial future. His quest for such knowledge grows out of no idle desire, although it is a quest that may often seem impracticable because, perhaps, its object lies beyond the means of a reasoned answer. Answers of various kind have indeed from time to time been offered, but only in the recent past has there emerged such knowledge as in its broad outline satisfies the demands of critical reason and of scientific fact. That answer goes by the name Evolution. It is set forth in this book reliably and simply by eminent authorities who have devoted their lives to a study of the evidence and to the work of making it more complete.

The creation of man is shown to have been a result, in

some respects the most striking result, of certain laws that hold throughout the animal and plant worlds. The more extended and more profound study of living things has revealed the manifold forms of animal life as one great series, in which the more complex are traceable by descent—or rather by ascent—from primitive simpler ancestral forms; and man is seen to have had his origin in a prehuman and subhuman animal stock, a stock which itself had in its time slowly attained to qualities and powers that made possible the attainment of man's own present estate. We can recognize in that estate a nature that relates us to much we might fain discard, and yet a nature that has been a passport for our further travel upward and has qualified us to achieve not only what man in the aggregate has achieved but what individual man at his best stands for. Thus, so far as face images forth mind, the reflection that from some simian grimace there has been evolved with the progress of time the smile of Mona Lisa is an exhortation to fortify man in his effort to gratify his yearning for higher things and for a yet more highly perfected future.

The creation of man perceived as a gradual and still operative evolutionary process, which, besides bringing him into existence is still moulding him and will not leave him where he is and as he is, bears broadly and profoundly on the interpretation of all human activities. This perception affords him new guidance in tracing to their origins his instincts, his emotions, his interests, and his reasoning power. In the light of this perception civilization and the history of civilization acquire fresh meanings; human society—its customs, its duties, and its growth—stand visible from a new angle and in truer perspective. There is incumbent, therefore, on every thinking man and woman, faced with the responsibilities of citizenship, an obligation to inform himself or herself, in at least some measure, of the nature and bearings of the great fact of evolution. Its principle is a part of established knowledge, acquaintance with which, by reason of the enlightment it sheds on life, each one of us, for our own sake and for the sake of others, should possess. To render help to those who seek such knowledge constitutes both the hope and the purpose of this book.

SIR CHARLES SCOTT SHERINGTON

THE TRAIL FOR ETERNITY

Mental and physical evolution plus Jesus - the trail for Eternity. Science tells us that all creation is made up of protons and neutrons. Together the protons and neutrons make up atoms, (a word which I'm sure the Jewish prophets mistook for Adam) which become different elements according to the amount of protons and neutrons in each atom... These different elements make up what we as beings see and are! We are evolving while made up of these same elements in zillions of body shapes. Our souls are a part of everything. Isahia said we are grass. Solomon felt we will see ourselves as beasts; "We all have one breath."

To only understand that the Holy Spirit is also One, with all included is really simple. Jesus told us, My Father, God, and I are One; Also He taught us that We are one with Him, Jesus... Clearly that make All Creation One.

It is clear to see that if we are One with Creation, and One with the Spirit then our total makes up God...

BILL HUTH

WHERE DO WE COME FROM?

Paul Gaugin painted cavemen with meditating thought - and then below his masterpiece he inscribed the solution not - only the ever unanswered riddle - five words that form the knot - of life's depth and it's beginning - and what was before - Where Do We Come From?

A biologist - Darwin by name - search of the answer to this question - became obsessed - forgetting fame - he sailed on the Beagle to a continent afar, observing the aboriginals - nature in the raw - for five long years he scribbled notes - then in climax two books he wrote - supposed to be the answer to man's own creation - Where Do We Come From?

The prophets of the Holy Books - the great men of time - gave themselves to reasoning the beginning of our vine - they sat in caves or under trees - some staring at the stars - in hopes to them would come the answer of life's dramatic mysteries - Where Do We Come From?

The child so young hears many tales - but truth he can understand - that for there to be a thousand answers - to such a simple thought - only proves his elders to be teaching what they learned not - Where Do We Come From?

BILL HUTH

"I believe that racial extractions and color hues and forms of worship become secondary to what men can do. The American public is not concerned with a first baseman's pigment as it is with the power of his swing, the dexterity of his slide, the gracefulness of his fielding or the speed of his legs."

BRANCH RICKEY

FORWARD

In this volume leading biologists of England and America, men distinguished in many special lines of research, are coöperating in a great endeavor to give the full meaning of the word "evolution." No word in any language at the present time is so comprehensive as this; few words are so misunderstood.

The original import of the word "evolution"—to unfold or to unroll, as a flower is unfolded—is too restricted, because, as theoretically presented in Lloyd Morgan's doctrine of emergence and as practically proved by palaeontologists in both the invertebrate and the vertebrate world since the time of Waagen, evolution is far more than the unfolding of something that already exists, as the germ develops and unfolds in the beauty of a rose; evolution is the incessant appearance of new qualities, new characters, new powers, new beauties, for which there is no antecedent in experience or no evident promise in the germ itself.

We almost feel the need of returning to the wonderfully adaptive language of the Greeks in an attempt to discover a new word or combination of words which shall better express all the many forms of activity Nature is now revealing far more clearly than when, in a relatively early and simple state of biologic knowledge, the word *evolution* was chosen as more appropriate than *mutation* or *transformisme.* If from Greek sources a new word could be borrowed or coined, it should certainly express the new principle that is implied in Lloyd Morgan's "emergence," in Bergson's *evolution créatrice,* in Osborn's "creative evolution," or in *"creation by evolution,"* the title of the present volume.

Are not new physical elements compounded by the simplification or complication of older physical elements, to give rise to new forms, but without the creation of new forces? Is there not invariably in the physical and material world antecedence and consequence, cause and effect? Are we not, therefore, facing in the biological world a new recognition of the order of Nature in the incessant creative, emergent evolution of new forms, of new characteristics, of new powers? Consequently the addition of new powers and new properties seems peculiarly distinctive of life.

Such questions, such problems, such contrasts as these show that Darwinism, broad and manifold in its implications as the term has become, is only one aspect of the whole evolution of life; there are many other and newer aspects, unknown to Darwin and not implied in the term "Darwinism," or even in the far more comprehensive term "evolution." As Einstein follows Newton, so some great philosopher of biology will follow Darwin, and the new biology of the future will be even more inspiring than the biology revealed by the many and able contributors to the present volume.

HENRY FAIRFIELD OSBORN

EVOLUTION OF GODLINESS

Mankind, the ultimate in receiving and transmitting both Spiritual and physical laws of nature on the planet earth, are being enlightened to the Spiritual and true facts concerning their physical being. Through the laws of natural selection, physical environment, and most of all Charles Darwin's working hypothesis--- physical evolution, the twentieth century generations are being offered at last a logical foundation for their being.

The uncontrollable urge to be at peace with oneself travels its course through the body of creation: the life Spirit we call God.

Life from the very beginning has contained its God.

To the tiny electron--the proton is God. Life giving Sun was God to Egypt. Movement created in the plants which surround us came through their constant desire to face what many call God.

For the one who claims to know and believe the principals of both physical and mental evolution it is impossible to separate himself from any Godly inspired law without first recognizing the effect it will have on himself as a functional part of creation. In this I speak both literally and factually.

Moses, who is buried in hypocritical versions of a God throwing molten stones from Heaven or parting the sea and purposely drowning the Egyptians said this, "who so sheddeth man's blood, by man shall his blood be shed." These eleven words, one sentence, stand as the truth and the very foundation for Charles Darwin's theory of evolution.

By taking this sentence which appears in Genesis, 9:6, anyone who can subscribe to the truth of physical evolution can connect himself with the Universe both physically and mentally--if I were to take your life, in reality I am taking my own: being a part of creation I am equally a part of you. Moses based his law -- "thou shalt not kill," on the very thought of all being brothers and an equal part of God.

EVOLUTION OF GODLINESS

If one man should by extra ordinary abilites kill every other person on earth he would in reality be taking his own life. For what other reason does a man exhibit his many talents than to please himself in the senses of others. For without creation there could be no God.

Solomon-the wise son of David spoke his feelings towards the question, "do beasts have the same Spirit as man?" The church of Solomon's time defended man, refusing to connect humankind with the beasts. "The beasts were created for food and labor for man."

Solomon, famous for his sound judgement, wisdom, intellect and decision informed his fellowmen in his book of Ecclesiastes 3:18-22 this:

"I said in mine heart concerning the estate of the sons of men, that God might manifest them, and that they might see that they themselves are beasts.

For that which befalleth the sons of men befalleth beasts; even one thing befalleth them: as the one dieth, so dieth the other; yea, they have all one breath; so that a man hath no preeminence above a beast; for all is vanity.

All go unto one place; all are of the dust, and all turn to dust again."

Who knoweth the spirit of man that goeth upward, and the spirit of the beast that goeth downward to the earth?

Wherefore I perceive that there is nothing better, than that a man should rejoice in his own works; for that is his portion: for who shall bring him to see what shall be after him?"

Think it over, "what shall be after him?" What type physical appearance will man contain in one million years?

Here hidden in a book who's very meaning is taught to be in complete discord with the teachings of evolution lies another great truth.

I am venturing on the great sea of life while endeavoring to prove and exhibit the benefits which can and do belong to you, me and all creation. It is only a matter of understanding and accepting, without doubt, the truth of Spiritual and physical evolution.

Our abode, Earth, rotates on its axis approximately every twenty-four hours. It circles the sun approximately every three hundred and sixty-five days. The Sun circles the center of the Milky Way on an orbit approximately sixty thousand light years broad. The Milky Way is approximately one hundred thousand light years broad and ten thousand light years thick. One light year being equivalent to 6,000,000,000,000 miles.

Astronomical Science also teaches us that there are over twenty-five million such Galaxies as ours and that there are over one hundred billion stars such as our Sun.

And then there is the great object for which all of this vast expanse was created; man!

Through continued and relentless study of the outer crust of the Planet Earth, geologists have proven beyond any doubt the existence of unbearable temperatures in the earth's nucleus at the present time. And that millions of years ago this temperature had enveloped the entire surface of our planet earth; It was a spinning molten ball of fire....A thing any "devil" would love.

The process of cooling resulted in a crust like surface on the sphere.

This was accompanied by severe expansions and contractions and the rushing of magma through giant cracks, and to the surface where it cooled into massive mountain peaks of past history.

The existing cracks in the Earth's surface today are both the results and reasons of our rumbling and disastrous earthquakes as the earth furthers its progress of cooling.

To study man and his part of this creation we must, in all fairness to ourselves, present this problem by starting at our present position and work towards the ever famous "beginning."

EVOLUTION OF GODLINESS

The most recent accomplishment of humankind is the ever increasing manner in which he enables himself to convey his thoughts which are conceived and improved by the intellect of all creation. Through the medium of conversation humanity has developed a means of taking the thoughts of one person and comparing them with those of another. This way allowing himself the benefit of being able to digest several thoughts and then conceive what will be good for himself as a member of society.

This means of conversation has been made even more beneficial through the written language. This language, both written and oral, has entitled man to greatly improve his cultural life.

To take these from humanity would be forcing him to return to the pre-caveman era. My momentary objective.

Etymologists have labored relentlessly. They have scrutinized the spoken and written languages of mankind throughout history in search of the definite pattern through which our present day languages were evolved.

The earliest recorded written language was supposedly conceived by the early Phoenicians. This was partly the result of vowel utterances and hieroglyphics. The latter being a mode of written symbol-like language of the Phoenicians and early Semetics of the far-eastern countries.

This period of time corresponded closely to the period in which Moses received through Godly inspiration the law, "an eye for an eye and a tooth for a tooth." The great "intellect" man at this time was acting out the part of savage. Sacrificial ceremonies with the offering of human lives to several different Gods are written history today.

Thought and knowledge at that period made it impossible to conceive the law, "if a man strike you on one cheek turn the other." Bless the person who takes your life was too much for the contemporary "beast"; to concede-it came nearly four thousand years later.

VOICE OF CREATION

Upon the walls of caverns throughout the European Continent archeologists have discovered thousands of examples of prehistoric art. Yet, with all of these there is still to be found one example of any type of written language other than picture writing. Simple deduction will provide the answer-if there was, at the time of these cave dwellers, a spoken language, it was limited to the symbols which were represented on their cavern walls.

This, a true fact, is a mental picture of our brute ancestors who lived as us, in search of truth. Unable to discuss and with very little intellect, these beasts were at that time the leaders of civilization. Their vocabulary, limited as it was, gave them a very weak insight as to their origination or destination.

And before this, factual as it may be, our ancestral background is engulfed in brute ignorance; beasts who also contained the desire to be at peace with oneself and limited to a vocabulary of grunts and groans-ah-eh-ee-oh-oo-the foundation for our present day vowels.

From this position we are forced to concede our great idol, man, to be masked with the physical appearance of the lowly beast; another, as Solomon might say, part of God.

We steal time from this beast once again.

By using the proof of finding sea shells on our highest mountain peaks we must surmise that at one time these mountains were buried beneath some type of ocean. Again geologists explain that many parts of the Earth have been underwater several times and that at one time the entire surface of the earth was covered by water.

The existence of air-breathing fish give the results of the great changes that have taken place in our ocean's sizes and depths. The Penguin, a bird that can not fly, another flattering result.

From water you were created and to water you shall return, were the thought for the day.

The near helplessness of the invertebrate-our next gift. And from this being we retrace our origin to the renown and very much under-estimated one-celled kingdom. Also an equal part of God.

The results of bisecting this one-cell will place us in the micro-universe of the elements which are made up of atoms and which in turn are the results of protons, electrons and neutrons.

And from here we enter the Ether Energy of Life.

Today, as it did billions of years ago, "Mother" Earth still carries a full supply of all of these parts of God: being a part of God herself. And for man to try as he might, never will he succeed in removing himself from the position of being, as much as he dreads it, an equal part of God with all creation; not something special.

The Earth was born a giant, flaming sphere of energy. Ever renewing its infinitesimal forms and regenerating life as it passes through eons of time on its own path of life.

As in all things planets, too, have their life periods to cope with. Eventually it shall pass into a stage of not being able to retain life as we know it and mankind shall be forced to escape to a new planet if it desires to carry forth civilization as we know it today.

If this is not done then the generations of relentless progress we have made will be lost and our accomplishments will have to be repeated as our Spirit moves ahead to a more and more peaceful life.

BILL HUTH

EVOLUTION—ITS MEANING

Evolution as Orderly Change

BY evolution, as the word is now used, we mean the universal process of orderly change. It includes cosmic changes in suns and planets and organic changes in living creatures, called organisms because they are made up of coöperating parts, or organs, which by fitting into one another constitute organization. And from the fact that all these changes—whether instantaneous, daily, yearly, or consuming centuries or æons, in the individual or in generations of individuals—are *orderly,* never random nor accidental, we derive our definition of evolution. Moreover, as this process occurs throughout all that we know, evolution becomes another name for Nature. Evolution, indeed, is Nature's way; thus all Nature study, if serious and thorough, must lead to the recognition of evolution. That Nature has her ways is the most visibly evident fact in all our experience, and such phrases as "blind force" have no real meaning.

Nevertheless, the forces and conditions which surround

suns and planets, or which mould mountains and seas, or which determine the formation of crystals or the accumulation of rocks, differ in certain ways from those which modify generations of life. We therefore usually treat orderly change in organized beings under a special head, that of organic evolution. For this a better term, bionomics, "life-ways," has been suggested by Professor Patrick Geddes, of Edinburgh.

The theory of organic evolution is, in brief, that in our world no living thing and no succession of living things remain exactly the same for any period of time, long or short; and furthermore, to repeat, that all change is *orderly,* never the result of accident or caprice or favoritism. In Huxley's words: "Nothing endures save the flow of energy and the rational order that pervades it."

As a science, organic evolution, or bionomics, comprises all that we know or that we may reasonably deduce from our actual knowledge of the history, development and divergence of living creatures on the earth. It involves the idea of the "transmutation" of species (or kinds of animals or plants) through natural causes (there are no others), their characteristics varying for cause, with time and with space. To one having a fair knowledge of the facts concerned no different working hypothesis is now conceivable; and a working hypothesis becomes a part of science when every rival hypothesis has ceased to work.

The evidence for organic evolution is cumulative. All creatures show evidences of evolutionary processes, which are revealed on every hand. Now that we have in some degree the clue to life and reproduction, every plant, every animal, every man, every institution appears (in its degree) not alone as an argument for but as a demonstration of evolution. Demonstrations precede logic and stand above argument. No other type of evidence, moreover, is so convincing as the cumulative one. The question we are considering is not one of logic but one of fact. Logic, with its specialized branch, mathematics, adds nothing to our knowledge; its function is to clarify assumptions already accepted.

Accepting the fact of orderly change—universal in so far as we can trace the relations of cause and effect and of natural sequence—we face a more difficult problem: How are the changes brought about? Here we no longer find unanimity of opinion, for in the myriad of facts at our disposal no single man can master their prodigious range and their diverse aspects. A forest is not the same to a lumberman as to a landscape gardener. A primrose in a greenhouse is not the same as one by the river's brim. "The harvest of the quiet eye" is not that which is garnered by the reaper. The microscope and the telescope yield knowledge from different angles. But the lesson of all science is that whatever takes place in nature is natural; not "supernatural." Indeed, to science "supernatural" is a meaningless word. It concerns either nothing at all or something not yet found out. We might say that the term "supernatural" can be applied only to a set of conceptions that are held by minds which have not learned that all facts of human experience are natural.

Much has been written as to the possible source of life in a lifeless world. It is easy to suppose some sort of "spontaneous generation" or "chemical transition." That supposition follows the line of least resistance; it is said by some to be a "logical necessity." Thus one sitting in his study may blithely construct "synthetic protoplasm" by "a juggling of words," or by a combination of ideas drawn from physics and chemistry. To state facts in simple terms, life appears only in connection with carbon, oxygen, hydrogen, and nitrogen bathed in light, heat, water, and air. So we all admit. But all life, so far as we know, starts from life, and every living being had some sort of living ancestry, moulded by the shifting and sifting of environment.

As to the origin of life on the earth we know nothing whatever. Speculation about it is more or less futile; indeed it may be mischievous, as when some particular unproved suggestion serves as a basis for further philosophical expansion. Science must stop *where the facts stop, or thereabout,* the limit of "thereabout" covering all legitimate diversions and excursions of philosophy.

Volumes have been devoted to the evidence of evolution, but their value depends on no single fact nor on hundreds of facts. The inevitable conclusion is that all the facts point the same way. All the evidence, whether drawn from comparative anatomy, embryology, physiology, or geographical distribution, from human institutions or from human history, brings us to the same result. All of it deals with the same truth as seen from a thousand different sides. All life has its roots in the past and its fruitage in the future. We must view the millions of kinds of living beings not as disconnected entities resulting from disconnected acts of creation, but as divergent twigs from the great parent tree of life. In a large sense, there is, as Parker observes, "only one kind of life in our world."

"What we mean by life is protoplasmic organization. Just what this is, we do not know. . . . It is continuous and has been continuous since the remote past and will continue indefinitely in the future. Vitality is the activity of the organization. Death is not of necessity the cessation of vitality; death occurs only with the disintegration of the machine. When this occurs with any single organism acting as trustee for the specific organization, there are myriads of other trustees which will carry that organization on and into the future." [1]

The reality of evolution in organic life once admitted, the next step must be to trace the details of its operation. For we recognize no "law of evolution" as working without regard to conditions. Evolution in vacuo is a philosophic fancy. Conclusions resting on analogies, or on the juggling of words, are not a part of science. "Living organisms," says Dr. Osborn, "differ from lifeless mechanisms, no matter how perfect, in being more or less self-adapting, self-reforming, self-perfecting, self-regenerating, self-modifying, self-resourceful, self-experimental, self-creative." In other words, they possess—

Individuality: No two organisms are exactly alike.

Irritability: The response to external stimulus, every organism being either swayed by influences bearing upon it or else reacting against them. Through evolution this response rises by degrees to tropism —the tendency to react in a definite manner—and to reflex action, with its specialized derivatives, instinct and intelligence.

Reproduction: The casting off of specialized cells, each one of which (usually united with its mate through amphimixis) initiates a new individual.

Metabolism: The wearing away of tissues and their replenishment by food derived from the substance of other organisms, or from water and from air.

Growth: The development in size and in specialization of the fertilized cell, which is followed by deterioration and death, except in one-celled organisms, where we have cell division instead of death.

[1] Calkins, G. N.

EVOLUTION—IT'S MEANING

Evolution: Modification of traits from generation to generation through internal and external factors.

The evolution of living beings or the transmutation of species is conditioned by at least four influences, always present and continually acting on every individual, animal or plant. These moulding factors are *heredity, variation, selection, segregation.* A species, as properly defined, is a kind of animal or plant which during countless generations has undergone these influences in the open, has thus run the gauntlet of life, and has endured. A sheltered form, watched over in a greenhouse or a breeding pen, is not a genuine species; to become one it must hold its own and survive outside, in the stress of Nature. "The origin of species" therefore concerns the coöperation of tendencies inherent in the organism, these being diverted, modified, or directed by obstacles without.

Inherent tendencies may be summed up as heredity and variation. Heredity is the conservative influence, which unifies groups, limiting divergence; variation is a force creating divergence. Variation results from a complex series of influences, the most obvious and apparently the most important being the biparental factor—that is, sex. External influences, acting on the traits that distinguish species, by serving as obstacles to the even flow of heredity, are selection and isolation. Selection destroys unadapted individuals, and often, through them, the types or species they represent. Isolation, with its consequent segregation, or prevention of mass-breeding, leads to the separation of minor groups from the original stock by barriers, mainly but not wholly geographical. Selection fits all types to their environment; it enforces adaptation on all living beings but does not divide them into species. Segregation is the final moulder of species. No sound discussion of species as they exist in nature can ignore geography.

Two general facts relating to the origin of species are often disregarded by those who are engaged in experimental work. The first fact, just referred to, concerns the relation of forms to geographical conditions; the second fact is that related species seldom differ in any survival trait or character by which one is better fitted to live than another. The "survival of the fittest" is a process that operates within the species rather than between one group as a whole and another group. Most species have one or more twins or geminates, which differ in minor features and do not inhabit the same region. This rule of geminate species, accepted by the ornithologist Dr. Joel A. Allen and called by him "Jordan's Law," was stated by the present writer in 1904, as follows:

"Given any species in any region, the nearest related species is not likely to be found in the same region, nor in a remote region, but in a neighboring district, separated from the first by a barrier of some sort, or at least by a belt of country the breadth of which gives the effect of a barrier."

Illustrations among plants, animals, races of men, and human speech appear on every hand. On either side of most barriers geminate species and subspecies (that is, species in the making) occur in every group of organisms, some so different as to require separate names, some barely distinguishable from their associates. Take those well-known birds the flickers, for instance. They belong to the genus *Colaptes,* a group of woodpeckers. On the east side of the Rocky Mountains we have the form called "yellow-hammer" (*Colaptes auratus*), with the shafts of its quills bright yellow. On the west side of the mountains we have the "red-shafted flicker" (*Colaptes cafer*), with the shafts of its quills bright red. These two species show also certain other slight differences, scarcely noticeable. On the other hand, the golden warbler, which ranges through the whole of the United States, migrating widely north and south as the seasons change, is all of one species, because everywhere it can breed freely with the mass of its kind.

Contrariwise, each island in the West Indies has its own peculiar species of warblers (Mniotiltidæ), whose migrations are not general but range simply between the mountains and the shore of the island. On the same principle and for the same reason, each island in the South Seas has its own peculiar dialect, each plainly derived, however, from the same ancestral language.

Again, each side of the isthmus of Panama, closed since Miocene time—two or more millions of years ago—has its geminate pairs of fishes, some six hundred in all, clearly defined, the distinctions being in traits as useless to the fish itself as to man. The temperate zone has its own series of forest trees, many of which are recognizable as geminates. The plane-tree, the elm, the elder, and the alder belt the earth, but with progressively changing species. In fact, in all groups the geminate relation becomes the rule, and a species absolutely isolated and unvarying is the exception.

These facts are too well-known to students of the geographical distribution of plants and animals to require elaboration here. It is therefore true, as already affirmed, that no theory of the origin of species can be sound if it fails to take geography into account.

Science and Faith

The Universe is with us. It is our Universe and we are part of it and have no alternative save to accept it as it is and as reverently as may be. The positive side of religion is the feeling of being at home in God's World. Whatever our conception of God the attitude remains. God's World is to us no alien land. It is our home and it has been the home of our ancestors for æons immeasurable, so that our life is fairly adjusted to it in all its details. And the more thoroughly and widely we become acquainted with its make-up the less sympathy we can feel with those who would "remould it nearer to the heart's desire."

Science is verified knowledge. Little by little, through processes of induction, we establish a basis of fact which, when stated in terms of human experience, becomes truth. No human truth, however, is absolutely without error. Yet though all truth must remain in some degree incomplete, given time it gains both momentum and accuracy. With such progress, error and false deduction fall off on every side, usually without any possibility of revival.

The history of science is marked by constant collision between tradition and discovery. In the majority of men, ideas are controlled by custom or by desire; hence arises the process, almost inescapable, called by Dr. Conklin "thinking wishly." Our observations and experiments may be quite objective; our thought, perhaps, is never altogether so. Anthropomorphic tendencies spread through our philosophy and through all the minor affairs of life and form a constant obstruction to the spread of knowledge.

Yet despite all this, and despite all forms of human credulity, science has forced the civilized world to acknowledge a good many things not hoped for and often not desired. We now understand, for instance, that the stars are not pinholes in the celestial floor, through which rain drips upon us; that the sunset is not lighted by the red flames of hell into which the sun daily sinks; that planets are not carried back and forth by angels; that light and heat both come from the sun; that the earth is not the immovable center of the universe; that in the universe we can "see no trace of a beginning nor prospect of an end," and that "time is as long as space is wide"; that the antipodes are really inhabited by real people; that 20,000 air breathing animals (outside of insects) could not foregather in pairs in the Ark; that fossils are relics of once-living creatures; that fossil shells are not evidence of the flood; that the Lord is not appeased by burnt offerings of lambs or of men; that lunatics are not possessed by devils nor yet struck by moonbeams; that the cure for scrofula is not found in the touch of a king; that no divinity indeed doth hedge a king, nor even the state; that a comet has its orbit and appears on its own business, not ours; that the penalty for wrong-doing is ours now and is within us; that ignorance and superstition are perilous guides for conduct; and that only Truth makes us free!

All our present conclusions concerning these matters and a thousand more are results of scientific research, not of religion as that word is commonly defined. To give one more example, however: Does any educated person now respect the dictum ascribed to Archbishop Ussher, who said, 200 years ago, that "Heaven and earth, center and circumference, were created all together at the same instant, with the clouds full of water, on October 24, 4004 B.C., at 9 o'clock in the morning?" And yet time was when to discredit this baseless pronouncement may have been held to cast one into outer darkness.

By the coöperation of observers and investigators much of the débris of our grandfather's science has already been cleared away, and with it necessarily the preconception of a special creation of the myriads of species of animals and plants and the assumed chasm between humanity and our lowly mammalian brothers. The collision of ideas which progressive discoveries of truth have occasioned is not, however, strictly speaking, a "warfare of religion and science." It is the inevitable struggle between tradition and knowledge, between conventional beliefs and new views demanded by new evidence. This conflict exists, not alone in church and state, but in the mind of every growing and forward-looking man.

The infinite expanse of the "unfathomed universe," its development through countless periods of time, the boundless range of its changes and the rational order that pervades it, all seem to demand an infinite intelligence behind its manifestations. That intelligence we cannot define, but of this we feel sure—it centers in no mere tribal god, nor one busy, man-fashion, with schemes and plans. Nor can it be one obsessed by human passions or jealousies. To thoughtful minds it becomes increasingly evident that the majestic mechanism of the universe and the perfect fitting of life to the earth on which it rests are no chance products of "fortuitous clashing of atoms." We know no cosmic results brought about by accident, happy or unhappy. It has been said that the attributes of humanity are merely traits of "complex carbon compounds." Even if true, this statement makes the facts no simpler, but far more complicated, by throwing on chemical reactions the brunt of the problems of life. So far as we can see, there is no "chaos" in the universe, nor was there ever any.

In the title of this symposium the word "creation" must be taken in its broadest sense as the aggregation of the intelligence and the energies which enter into the development of the Universe. Is not "creation by evolution" a far more exalted conception than any creation by fiat imagined of old? And does it not reveal a Godhead infinitely worthy of obedience and adoration?

DAVID STARR JORDAN

A sacred kinship I would not forego
Binds me to all that breathes; . . .
I am the child of earth and air and sea.
My lullaby by hoarse Silurian storms
Was chanted, and through endless changing forms
Of tree and bird and beast unceasingly
The toiling ages wrought to fashion me.

Lo! these large ancestors have left a breath
Of their great souls in mine, defying death
And change. I grow and blossom as the tree,
And ever feel deep-delving earthly roots
Binding me closely to the common clay;
Yet with its airy impulse upward shoots
My soul into the realms of light and day!

Hjalmar Hjorth Boyesen

LIFE'S WAY OF EXISTENCE

While Darwin was on his voyage aboard the Beagle, the Ship was sailing along the coast of South America. They stopped at a group of small Islands and visited with the natives. While they were there they learned that when things get dearth, even though they had dogs around the camp, the natives killed and ate the older women of the tribe.

Life has many ways of displaying it's means of selecting it's method of existence.

BILL HUTH

WHY WE MUST BE EVOLUTIONISTS

Evidences of Evolution

We use the familiar phrase "evidences of evolution" with some misgiving, because it does not suggest the right way of looking at the question. Evolution means a way of Becoming. Just as it is certain that all the many races of domesticated pigeon are descended from the wild rock dove, so, it is argued, have all the different kinds of wild animals and wild plants descended from ancestors that were on the whole somewhat simpler, and these from simpler ancestors still, and so back and back until we come to the first living creatures, whose origin is all in the mist. Evolution just means that the present is the child of the past and the parent of the future.

But it is not possible to prove this conclusion in an absolutely rigorous way. We can, indeed, see evolution going on now, but we cannot, so to speak, reverse the world-film and see precisely what took place long ago. The records in the rocks do clearly reveal what happened in the past, even millions of years ago, but not in so clear and so detailed a way as the developing egg of a hen reveals the gradual rise and progress of the chick.

Although we do not know of any competent biologist

to-day, however skeptical and inquiring he may be, who has any doubt as to the fact of organic evolution, yet no one would assert that it can be demonstrated as one might demonstrate the law of gravitation, or the conservation of matter and energy, or the development of a chick out of a drop of living matter on the top of the yolk of the egg. But how can a conclusion be accepted without hesitation if it is not rigorously demonstrable? The answer is that the evolution-idea is a master key that opens all locks into which we can fit it, and that we do not know of a single fact that can be said to be in any way contradictory. Like Wisdom, the evolution-idea is justified of its children.

A great zoölogist once said that he was willing to stake the validity of the evolution-idea on the evidence afforded by butterflies, and he was quite right. Any fact about an animal or a plant may be an evidence of evolution when we know enough about it. What makes the general idea of evolution convincing is its satisfactoriness in interpretation. It is always borne out by the facts. We repeat the phrase "the general idea of organic evolution" because this must be distinguished from any particular theory in regard to the factors that have operated in the process. In regard to the *factors* or *causes* of evolution there is, and there may well be, difference of opinion among naturalists, for the inquiry is as young as it is difficult; but it is unfair and confused to use this admission of uncertainty as to *causes* as if it implied any hesitation in regard to the *fact* of an age-long evolutionary process in which many of the highly finished and very perfect types of animals are shown by the rock record to be preceded by a succession of animals in less finished stages.

There is eloquence in the evidence from the rock record. As ages passed there was a gradual emergence of finer and nobler forms of life. Among back-boned animals the first were the fishes. These led to the amphibians, and these were succeeded by reptiles. Later there arose birds and mammals. Throughout the ages, life has been slowly creeping upward. Detailed pedigrees are disclosed in the rocks, some of them with marvellous perfection, as in the evolution of horses and elephants, camels and crocodiles. For some animals, such as fresh-water snails and marine cuttlefishes, there is an almost perfect succession of fossils, forming a chain in which link 10 is very different from link 1, yet just a little different from link 9, as link 2 is a little different from link 1. For such animals we can almost see evolution anciently at work!

The geographical evidences are also endless. If the present state of affairs is not the outcome of a natural process of evolution, why should the fauna of oceanic islands be restricted to those animals which can be accounted for by transport over the sea by currents and by winds, or on the feet of birds? Thus there are no amphibians on oceanic islands, because few amphibians can endure salt water.

The inhospitable Galapagos Islands are said to be the submerged tops of cold volcanoes, which belong to an ancient peninsula that became first an island and then an archipelago. They have a peculiar fauna, which includes the famous giant tortoises. There are ten different kinds of giant tortoise on ten different islands, and those that are on the islands that are farthest apart are most unlike. There are five different kinds in different parts of the largest island, which is called Albemarle. Now if we consider thoughtfully these facts what can we find them to mean except that isolated groups of one ancient stock of the original peninsula have varied slightly on one or another island and that the isolation prevented any pooling or blending of the new forms? For these large tortoises cannot swim. On Albemarle Island the isolation is probably topographic; it is due to barriers formed by the rugged volcanic surface. When Darwin, as a young man, visited these islands during the voyage of the *Beagle* he was greatly struck by the fact that each island seemed to have its own kind or species of giant tortoise, and he tells us that he felt himself "brought near to the very act of creation." This was one of the experiences that made Darwin an evolutionist.

But think also of the anatomical evidence. It is interesting to compare a number of fore limbs—our own arm, a bat's wing, a whale's flipper, a horse's fore leg, a bird's wing, a turtle's paddle, a frog's small arm and a giant giraffe's at the other extreme. They are very different, and yet when we scrutinise them we find the same fundamental bones and muscles and blood-vessels and nerves. "How inexplicable," Darwin said, "is the similar pattern of the hand of man, the foot of a dog, the wing of a bat, the flipper of a seal, on the doctrine of independent acts of creation! How simply explained on the principle of the natural selection of successive slight variations in the diverging descendants from a single progenitor." Few zoölogists of today would use Darwin's words "how simply explained," for we are aware of factors he did not know of, and some of the factors he believed in very strongly are not unanimously accredited today. But all would agree that the evolution-idea illumines the deep identities, amid great superficial diversities, that are disclosed when we consider, let us say, the classes of backboned animals.

Another anatomical argument is to be found in the frequent occurrence of vestigial structures in animals and in ourselves. Useless dwindled relics of the hind limbs of a whale are found buried deep below the surface. In the inner corner of our eye there is just a trace of what is called the third eyelid, a structure that is strongly developed and readily seen in most mammals, as well as in birds and reptiles. It serves to clean the front of the eye; but although it is big enough to do this in most mammals and birds it is a mere relic in man. Take another example: behind the eye of the skate—a familiar flat fish—there is a large hole called the "spiracle." It serves for the incoming of the "breathing water," which washes the gills and passes out by the five pairs of gill-clefts on the under surface. But if we peer into this very useful breathing-hole or spiracle we see a minute comb-like structure, which is the dwindling useless relic of a gill. The cleft or spiracle is of indispensable use to the skate, but the relic or vestigial gill inside the spiracle is of no use at all. Yet it tells us that a spiracle was evolved from a gill-bearing gill-cleft.

One of the most remarkable sets of facts about living creatures—plants as well as animals—is that old structures become transformed into things very new. The poet Goethe helped to make the great discovery that the parts of a flower—sepals, petals, stamens, and carpels—are just four whorls of transfigured leaves, the stamens and carpels being spore-bearing leaves. We sometimes see the whole flower of a flowering plant that has become too vegetative "go back" and become a tuft of green leaves; and it is an unforgettable lesson to pull the flower of the white water lily to pieces and to find that the green sepals pass gradually into white petals, and these gradually into yellow stamens.

Similar lessons are taught by animals. What is the sting of a bee but a transformed egg-laying organ or ovipositor (therefore never found in drones), and what is an ovipositor but a transformed pair of limbs? The elephant's trunk was a great novelty in its way, but it is just a very long nose with an additional piece due to a pulling out of the upper lip. This is the evolutionary way!

We live in what has sometimes been called the "age of insects," for of these there are more than a quarter of a million different kinds. Now there must be some meaning in the fact that these can be classified in an orderly way; that one can for many kinds make plausible "genealogical trees." Often one species, with its varieties, seems to grade into another. In many parts of the animal kingdom there are types that link great classes together. Thus the old-fashioned *Peripatus* type, a little creature somewhat like a permanent caterpillar, has some worm characters and some centipede characters. It is to some extent a connecting link. The oldest known bird, a fossil beautifully preserved in lithographic stone of Jurassic age, has numerous reptilian features, such as teeth in both jaws, a long lizard-like tail, a half-made wing, and abdominal ribs. Yet it was a genuine feathered bird! And this fossil is unexplainable unless we recognize the fact that this bird had reptilian ancestors.

Very striking, again, are the embryological facts which show that the development of the individual is like a condensed recapitulation of the probable evolution of the race. An embryo bird is for some days almost indistinguishable from an embryo reptile; they progress along the same highroad together; but soon there comes a parting of the ways and each goes off on its own path. The gill-slits of fishes and tadpoles—the slits through which the water used in breathing passes—are persistent in all the embryos of reptiles, birds, and mammals, though in these higher back-boned animals they have nothing to do with respiration. All of them are merely transient passages except the one that becomes the "eustachian tube," which leads from the ear to the back of the mouth. They are straws which show how the evolutionary wind has blown. In a great many ways the individual animal climbs up its own genealogical tree, but we must be careful not to think that an embryo mammal is at an early stage of its development like a little fish, as some writers have carelessly said. Each living creature, from the very first stage of its development, is itself and no other; and though the tadpole of a frog has for some weeks certain features like that of a fish, especially a larval mud-fish, it is an amphibian from first to last. The embryo is the *memory* of a fish or of a reptile-like ancestor. There is no doubt that the hand of the past is upon the present, living and working; and this is evolution.

Many living creatures today are like ever-changing fountains; they are continually giving rise to something new. The beautiful evening primrose (*Oenothera*) and the American fruit-fly (*Drosophila*) are notable examples of changeful types; they are always giving birth to novelties or new forms, technically called "variations" or "mutations"; but the fact of variability is widespread.

In some forms the breeder or the cultivator is able to provoke great changes, for instance, by altering surroundings and food; but he usually has to wait for what the natural fountain of change supplies. This has been our experience with the domesticated animals and cultivated plants that interested Darwin so much. All the domestic pigeons have been derived, under man's care, from the blue rock dove; and there is strong evidence that the multitudinous breeds of poultry are all descended from the Indian jungle-fowl. What Darwin said was this: If man can fix and foster this and that novelty and make it the basis of a true-breeding race, and all in a comparatively short time, what may Nature not have accomplished in an unthinkably long time? And when it was objected: But what is there in Nature corresponding to Man the Breeder, his characteristically Darwinian answer was that the Struggle for Existence implied a process of sifting, which he called Natural Selection. Testing all things and holding fast that which is good or fit: that has been the evolutionary method!

These few examples should make plain the nature of the argument for evolution. It is what is called a cumulative argument. All the lines of facts meet in the same conclusion—the present is the child of the past. There is no conflicting evidence; every new discovery points in the same direction. On many sides we find striking facts, which become luminous when we see them in the light of the evolution-idea. But without that light they are worse than puzzling. All the facts conspire toward the conclusion that animate nature has come to be as it is by a continuous natural process, comparable to that which we can study in the history of domesticated animals and cultivated plants. But we do not give a satisfying account of what has taken place until we can state all the factors that have operated, and that is the subject of the much-debated detailed theories of evolution, like Darwinism and Lamarckism. And even if we were agreed about the factors we should still have to inquire into the meaning or significance of the whole. But that is a religious question.

An Enriching Outlook

Another great reason why we must be evolutionists will come as a surprise to some people. The evolutionist outlook is one that lightens the eyes and enriches us. We are impoverishing ourselves if we shut out the light of evolution. Let us consider three points only.

1. The evolution-idea gives the world of animate nature a new unity. All living creatures are part and parcel of a

great system that has moved sublimely from less to more. All animals are blood-relations; there is kinship throughout animate nature.

2. It is indeed a sublime picture that the evolutionist discloses—a picture of an advancement of life by continuous natural stages, without haste, yet without rest. No doubt there have been blind alleys, side-tracks, lost races, parasitisms, and retrogressions, but *on the whole* there has been something like what man calls progress. If that word is too "human" we must invent another.

3. One of the greatest facts of organic evolution—a fact so great that it is often not realised at all—is that there has been not merely an increase in complexity but a growing dominance of mind in life. Animals have grown in intelligence, in mastery of their environment, in fine feeling, in kin-sympathy, in freedom, and in what we may call the higher satifications.

No evolutionist believes that man sprang from any living kind of ape, yet none can hesitate to believe in his emergence —"a new creation"—from a stock common to the anthropoid apes and to the early "tentative men." Long ago there was a parting of the ways—it could not be less than a million years ago: the anthropoids remained arboreal and the ancestors of the men we know became terrestrial. So far as we can judge from links that are certainly not missing, but always increasing in number, there were for long ages only tentative men like *Pithecanthropus* the Erect, in Java, and *Eoanthropus,* the Piltdown man of the Sussex Weald. Even these were rather collateral offshoots than beings on the main line of man's ancestry. They were Hominids, but not yet Homo. What trials and siftings there seem to have been before there appeared "the man-child glorious!" Doubtless some great brain change led to clearer self-consciousness, to language, to a power of forming general ideas, to greater uprightness of body and mind; and it is very important to realise that a steady advance in brain development, on a line different from that of other mammals, is discernible in the very first monkeyish animals. Man stands apart and is in important ways unique, but he was not an abruptly created novelty. That is not the way in which evolution works. Man, at his best, is a flower on a shoot that has very deep roots. What the evolutionist discloses is man's solidarity, his kinship, with the rest of creation. And the encouragement we find in this disclosure is twofold. In the first place, though we inherit some coarse strands from pre-human pedigree, it is an *ascent,* not a *descent* that we see behind us. In the second place, the evolutionist world is congruent with religious interpretation. It is a world in which the religious man can breathe freely. To take one example: there are great trends discernible in organic evolution, and the greatest of these are toward health and beauty: toward the love of mates, parental care, and family affection; toward self-subordination and kin-sympathy; toward clear-headedness and healthy-mindedness; and *the momentum of these trends is with us at our best.* And evolution, with these great trends, is going on: Who shall set it limits?

J. ARTHUR THOMSON

"Invisible, impalpable foreces steaming around us and through us; perpetual change and transfromation on every hand; every day a day of creation, every night a revelation of unspeakable grandeur; suns and systems forming in the cyclones of stardust; the whole starry host of heaven flowing like a meadow brook."

JOHN BURROUGHS

VOICE OF CREATION

PROGRESSIVE EVOLUTION

"Progressive evolution is the universal plan. Everything which we meet in the world around us, matter and mind, every individual and all congregated masses, begin their course as germs and unfold in slow progression . . .The faculties of all intelligent creation, all that you call mind, all that you call heart, are framed for an interminable series of evolutions. . . . It is not mainly the mould of this mighty frame of things which establishes it, it is the fact that creation is eternally unfolding new resources and presenting itself under successive and amazing combinations of which no creature in the universe had imagined it capable."

JAMES McCLOUD

A Presbyterian Minister of Lexington, Kentucky, in 1818, when Darwin was a small boy.

IT'S ME

I, am concepted in black. I, am concepted in yellow. I, am concepted in white. I, am concepted in red. I, am concepted in the mixture of all colors. All colors are Mine.

I, am concepted with two arms and two legs. I, am concepted with no arms and forty legs. I, am concepted with four legs and no arms. I, am concepted with no arms and no legs. Wings and fins are also Mine.

I, am concepted as grass. I, am concepted as the vine. I, am concepted as the flower. The tree, too, is made of Me.

I, am concepted as the sun. I, am concepted as the atom. I, am concepted as the earth. All of Me is the Universe.

BILL HUTH

CAN WE SEE EVOLUTION OCCURRING?

The doctrine of organic evolution is the doctrine that animals and plants are slowly transforming, producing new kinds; that they have done this in the past and are continuing to do it now. It does not deal with something transcendental, something metaphysical; it deals with processes as real as the running of a stream or the growth of a tree. Organic evolution, then, is a physiological process, like the digestion of food; it is something that is occurring at all times, including the present. The doctrine of organic evolution means simply that if you lived long enough you would see organisms begin as simple creatures, change shape and structure as a growing plant does, become diverse, transform repeatedly, until from one or a few types many would be produced. You would get dissolving views of amoeba transforming to creatures having more definite structures and greater complexity; of *Hipparion* becoming a horse; of an ape-like creature becoming a man.

But no one can live long enough to see all that. The process is too slow. Within the life of a man, or of many successive men, very little alteration can be expected. Yet men have detected and measured other slow things. The earth's pole swings about in a small circle with a movement so slow that it requires 25,000 years to go once around the circle; yet men have discovered this fact and have measured the present rate of the motion. The fixed stars are not in fact fixed in their relation to one another; slight changes of position occur, some of them requiring centuries for detection; yet men have detected them. Certain radioactive substances disintegrate so slowly that it requires millions of years for a given portion to transform, yet the changes have been detected and their rate has been measured. If we can detect these things why should we not be able to detect—to catch in progress—the changes that we call evolution? We cannot directly see the growth of a tree, but by taking photographs at intervals and running them through a moving picture machine we can see the tree grow, and we can determine how its growth occurs. Ought we not to be able to get some sort of a moving picture of evolutionary change?

The task is bound to be difficult. The process of evolution is complex. Evolutionary changes move in many different directions. Some organisms degenerate; others grow more complex and become adapted to more varied conditions; still others change hardly at all. The process cannot be uniform; it produces diversity, not simplicity. What sort of changes should we expect to find if we could watch certain organisms closely enough to see the evolutionary changes that occur in them during a human lifetime?

We should have to study some organism that produces many generations while man passes through but one. Fortunately, there are many such organisms—creatures that produce a new generation every day or every few days. We should have to begin with a single individual and follow its offspring for many generations, obtaining great numbers of descendants. According to the theory of evolution, slight changes would occur as the generations pass. These changes would not all be purely transient, disappearing with the next generation; some of them would be hereditary; they would persist in later generations. The descendants of the original single individual would become diverse—hereditarily diverse. From a single individual, from a single race, we should thus, after the passage of many generations, get many races that would be hereditarily diverse.

In a human lifetime or in many human lifetimes we could not expect these changes to be great. Geological time is enormously long and evolution is prodigiously slow. The doctrine of evolution would therefore not lead us to expect to see widely diverse creatures produced. The popular demand that we should see a cat, or the offspring of a cat, transformed into a dog, or an amoeba into a vertebrate, is not in accord with the doctrine of evolution. We cannot expect in a lifetime to see new "species" produced. All that the doctrine of evolution leads us to expect is that there should appear slight hereditary changes, so that from a single race there are produced a number of hereditarily diverse races, differing slightly.

Do we find this? Studies of this sort have been made of a number of organisms. What was found in such a study made by the present writer may be set forth as a type.

It is common to suggest that amoeba or some amoeba-like creature is the original stock from which animals descended; "from amoeba to man" is a common phrase. It is of interest to examine amoebas from this point of view. Are amoebas still transforming, producing other kinds of animals? Some of the amoebas are naked and formless, so that the detection of any slight hereditary changes would be almost impossible. Others have shells of definite form and structure, furnishing excellent opportunity for the detection of hereditary alterations. These shelled amoebas, though they closely resemble the naked ones, are designated by other names. One called *Diffiugia corona* (Fig. 1) was selected for observation and breeding. It is a miscroscopic creature about 1-150th of an inch in diameter.

These creatures multiply for long periods without any sexual process; that is, each individual divides into halves, and each half then develops into a complete cell, which is later in turn subjected to the same dividing process. Any individual is therefore the offspring of but a single parent; not of two parents, as in the higher animals. The method of reproduction in *Difflugia* is shown in Figure 2. A new generation is produced about every two to four days, so that in the course of a year or two many generations may be followed through thousands of descendants produced from one individual.

Do these thousands of descendants all remain hereditarily alike? Or do they gradually and slowly diverge, becoming hereditarily different, as the doctrine of evolution sets forth?

This was studied by allowing a single individual to reproduce for many generations, until it had produced thousands of offspring. In the early generations of such an experiment, hereditary changes cannot be detected. The offspring often differ from the parents in certain respects, but most of these differences appear not to be inherited. The next generation shows similar differences, but as the generations increase in number we find that certain diversities accumulate and become hereditary. In some descendants the spines become longer; in others they remain shorter. In some the bodies are larger; in others they are smaller. Different combinations of size of bodies and of length of spine appear. These differences are inherited. In time from the original single individual a number of diverse stocks have

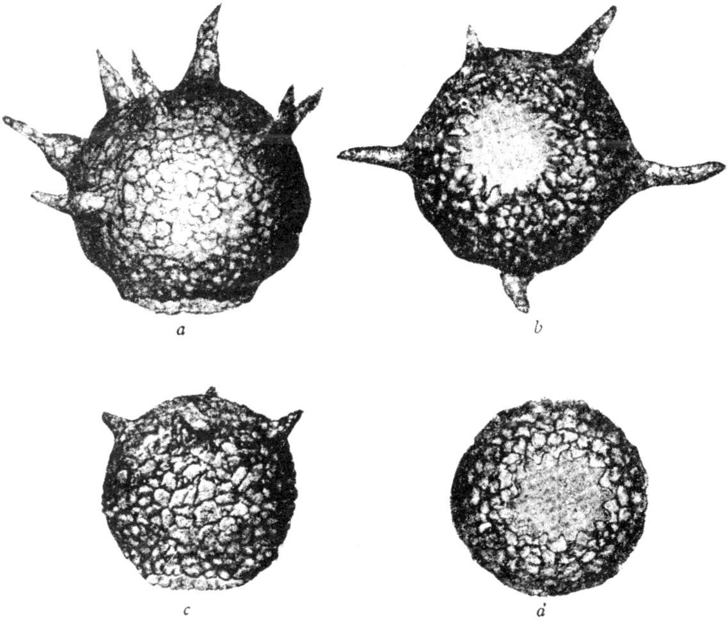

FIG. 1.—Microscopic views of a shelled amœba.

A medium-sized amœba is over a million million times smaller than the smallest mammal.

Shells of two diverse races of *Difflugia corona,* showing characteristics that become hereditarily altered as generations pass. The individual shown at *a* and *b* (a side view and a view looking at the mouth) bears about seven long spines and has sixteen teeth around the mouth; the individual shown at *c* and *d* has four small spines and fourteen teeth.

been formed (Fig. 3, *B* to *F*.). These five sets are like different branches of a tree, all coming from one trunk. They are separated by many generations from the original parent A. During the passage of these generations the different branches have become permanently diverse. Each differs from the other hereditarily. Even when all are living

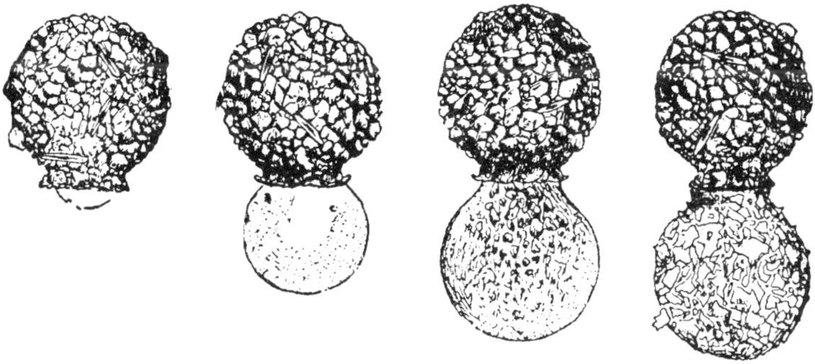

A *B* *C* *D*

FIG. 2.—Method of reproduction of *Difflugia* (after Verworn).

The parent *A* consists of a mass of protoplasm, covered by a rounded shell made of sand grains cemented together. This shell has an enlarged opening (below, in the figures). During its life the parent creeps about at the bottom of pools. It takes up many sand grains, which it stores within its body. At reproduction the protoplasm of the parent swells and projects from the mouth of the shell *A*. This projecting part enlarges and takes a form similar to that of its parent (at *B* and *C*). The nucleus of the parent divides and one-half of it passes into the projecting portion. The sand grains within the parent body also pass out into the projecting mass, come to its surface, and spread over it (*C* and *D*). They are embedded in a fluid secretion which now turns hard, forming a shell like that of the parent. The two shells are in contact at their mouths. Now the mass of protoplasm divides into two individuals, which separate, one retaining the old shell (above in the figures); the other having the new shell. Later, each individual repeats this process, producing another generation.

under the same conditions the stocks remain diverse for generation after generation. From the original single stock several hereditarily diverse stocks have been produced. Each set or race included a large number of individuals, all showing the characteristics illustrated by their representatives in the figure. A single stock, derived by fission from a single parent, has gradually diversified itself into many stocks that are hereditarily different.

What the doctrine of evolution asserts is therefore true for *Difflugia.* It does gradually transform and produce new races. If this is what evolution means, we have here seen evolution occurring.

A number of other lower organisms have been studied in a similar way, and with similar results. They do not remain entirely constant. Although the process is excessively slow, they gradually transform into hereditarily diverse races, in accordance with the doctrine of evolution.

To observe such changes in higher animals and plants is much more difficult. Each generation requires a longer time; in a human life few can be observed. But a greater difficulty lies in the fact that most of the higher organisms reproduce from two parents. The two parents always differ in their hereditary constitution, so that the offspring are usually a combination of two hereditarily diverse stocks. In forming that combination, each parent loses half of its genes—that is, half of the thousand different chemicals on which depend the way it develops and its later characteristics. The remaining halves from the two parents then unite to form a new combination of genes, from which the offspring develops. For every single offspring the process is repeated, but in each case it is a different set of genes that is lost from each parent, a different set that remains. Consequently through the union of the two remaining halves there is in every case a new and diverse combination of the genes produced; so that every one of the offspring of a pair of parents differs in its hereditary constitution from every other one;

CAN WE SEE EVOLUTION OCCURRING?

A, Original parent from which *B, C, D, E, F* are derived.

B, Individuals small, with small spines.

C, Bodies somewhat smaller, with larger spines.

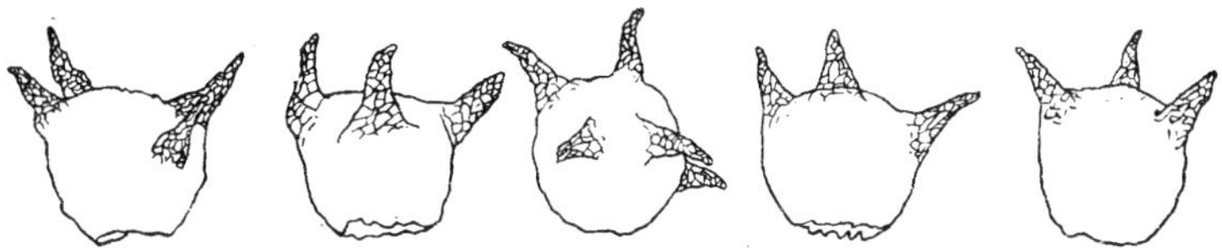

D, Spines still larger; bodies about the same size as *C.*

E, Animals larger than *A, B,* or *C* and spines larger than in *C.*

F, Bodies larger than others, but spines small.

FIG. 3.—Five different races of *Difflugia corona* derived from a single stock.

as well as from both of the parents.[1] These differences show themselves in the characteristics of the developed individuals, in thousands of diverse ways, some very marked, some extremely inconspicuous. It becomes therefore extremely difficult to distinguish differences produced in this way—by recombinations of genes in biparental reproduction—from differences that are steps in evolution. In most higher organisms this is indeed at the present time impossible.

Yet in certain higher organisms these difficulties have been overcome. By study, continued for years, the hereditary constitutions of the parents are thoroughly learned, so that the results of their combination are known. In such organisms the hereditary constitution does change at times, irrespective of the recombinations due to the union of two parents. The changes are infrequent. Yet in such an animal as the fruit-fly (*Drosophila*), studied by Morgan and his disciples, so great has been the number of individuals and of generations minutely studied that literally hundreds of different alterations in hereditary characters have been observed. *Drosophila* has given rise to hundreds of new stocks, which differ permanently from the original one. Some of the changes are strongly marked, as when red-eyed animals suddenly produce a white-eyed race, or when long-winged creatures suddenly produce a race that is permanently without wings. These very marked changes were naturally the first ones observed, so that for a long time it was believed that all evolutionary changes were large leaps, saltations. But since acquaintance with the animals has become more minute it has been discovered that extremely slight, almost imperceptible, changes in hereditary characters are much more

[1] For details as to this process of recombination of genes in reproduction from two parents, see any modern text-book of genetics; e.g., T. H. Morgan's *The Physical Basis of Heredity.*

common than large ones. Dozens of different faint gradations in the color of the eye have appeared. Physiological changes so slight that they can be perceived only after long experimental study have been noted in great numbers. Every feature of the animal has thus become modified in many different ways. Hundreds of diverse races of *Drosophila* have taken origin from the original one; and many of these are more diverse than what have been called different species in this genus.

We do not yet understand the causes of these changes; we do not know how they are produced. An immense deal remains to be learned about them. But our ignorance must not be allowed to obscure the great, the essential fact that appears in these attempts to see evolution in progress—the fact of actual change. Remember that there are two opposite doctrines. One holds that the constitution of organisms is permanent; that they were created as they are and do not change. The other, the doctrine of evolution, holds that the hereditary constitution slowly changes as generations pass; that a single race differentiates in the course of time into diverse ones; that from one stock many are produced. The critical observations that have been made on these minute living organisms through the passage of generations substantiates this theory; they do change and differentiate into diverse races as generations pass. The facts observed are what the doctrine of evolution demands, not what the opposed theory demands.

HENRY SPENCER JENNINGS

VOICE OF CREATION

EVIDENCE OF EVOLUTION

"The evidence of evolution had been read in the rocks and the structures of plants and animals, but under the miscroscope Dr. Jennings is able to follow *evolution* not as a theory but as *a thing that is actually taking place.*

"Intensified study reveals that the hereditary characteristics do become changed by external conditions. Through such diversities, continuing for great numbers of generations, single stocks, uniform in their hereditary characteristics, gradually differentiate into many faintly differing hereditary features.

"In higher organisms the state of knowledge of this point appears less satisfactory. But the evidence, so far as it goes, indicates that processes here are in agreement with those in lower organisms.

"The organisms whose bodies are condensed into a single cell have, too, a life condensed into a few hours. They present a wonderful opportunity for solving in a brief period some of the deeper problems of life.

"In a watch glass on our table we may in a week see generations come and go. We may follow in successive generations the struggle for existence and the results of natural selection.

"In a few days we may see the birth, babyhood, youth, and age of individuals and their replacement by descendants. We may study the inheritance of parental traits by the new generation or the appearance of new traits. We may observe how the population changes with the passage of ages—all while we wait for one of the changes of the moon.

"What have the simple organisms to teach us on youth and age, death and rejuvenescence, heredity, variation, evolution?"

DR. JENNINGS

THE LAW OF EVOLUTION

"In scientific deductions one single divergency suffices to demolish the structure on which they are based."

Evolution is not a force, but a process; not a cause, but a law.

LORD MORLEY

TOGETHER WE ARE

Jesus said, "you and I are One." The He added, "My Father and I are One." If this is true then Jesus, our Father and you and I are One. So what's there to worry about?

BILL HUTH

VESTIGIAL ORGANS

THE body of an animal, like a piece of machinery, is made up of numerous parts that work together toward a common end. In an animal these parts are the organs, and each organ commonly has a definite use. Thus in man the eye, the hand, and the heart are three organs, the first for vision, the second for prehension, and the third for the propulsion of the blood. As in the machine, so in the animal, some parts are more important than others.

Just as there may be superfluous wheels and belts in a machine, so there may be organs in an animal that are not essential. A man may lose an arm or a leg and still live. Paired organs may be reduced by the removal of one. Thus a surgeon does not hesitate to excise a kidney provided its mate can be left intact. After such an operation the organ left behind usually enlarges and acts for two. Some single organs, such as the spleen, or even the stomach, may be removed without causing death. After its removal the function it ordinarily performs is taken over by other organs, and the life of the individual continues.

But in addition to organs of the kind just described there are others whose loss is followed by death. None of the higher animals can survive the loss of the heart, and every such animal requires at least the most of one lung and of one kidney. Even such inconspicuous organs as the adrenal glands, small bodies lodged in the fat near the kidneys, cannot be removed without causing death. Organs that are thus absolutely essential to the continuance of life are called vital organs and stand in strong contrast to others whose functions are not so essential. Different organs thus vary

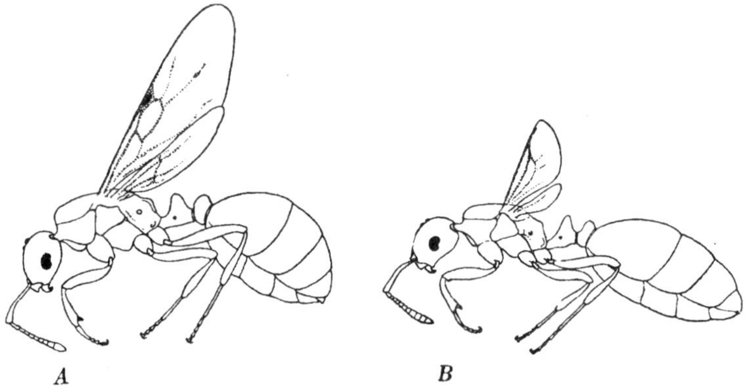

FIG. 1.—*A*, Female of an ant (*Monomorium rothsteini*) with normal wings; *B*, Female of an ant (*Monomorium subapterum*) with vestigial wings.

greatly in that some are essential and others are not essential to life.

Vestigial organs are those that are quite useless. Such organs are of course not necessary to their possessors. It is not always easy to prove that a given organ is vestigial. As might be inferred from what has been said, such proof requires that a given organ be not only removable without detriment from the animal of which it is a part but that it be shown to be without function. Conclusive evidence on this point is not always easy to obtain, for our ignorance may at times leave us in doubt as to the presence or absence of function in a given part. Notwithstanding this difficulty we know that many organs are truly vestigial, and such organs may be found in animals of almost any group.

Most insects are capable of flight and possess to this end a pair or, more commonly, two pairs of wings. Nevertheless many insects have wings that are entirely useless. Thus the male of the gipsy moth has well-developed wings and flies as other moths do; but the female, though she has fully formed wings, makes no use of them. When she emerges from her cocoon she creeps a short distance away and deposits

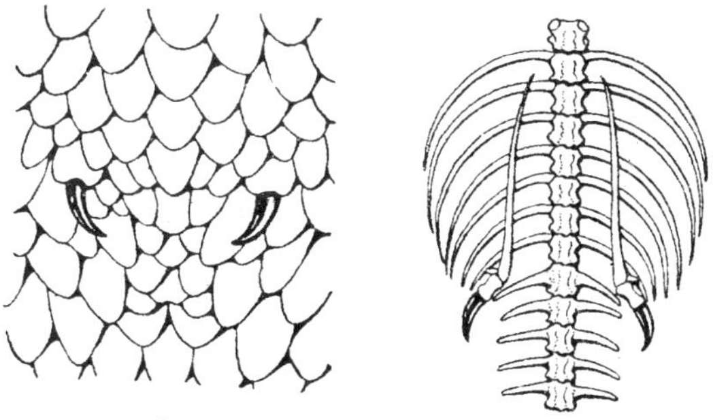

A *B*

FIG. 2.—*A,* Claws at the sides of the vent of a python, representing vestigial hind legs; *B,* Skeleton supporting the claws. After Romanes.

her eggs, but without flight. Her wings are functionless and in that sense vestigial. In many other insects the wings are not only useless but are relatively small. Examples are seen in certain chalcids, small, almost microscopic wasp-like creatures that are often parasitic in other insects. Wheeler has recently described an Australian ant, *Monomorium subapterum* (Fig. 1), in which the wings of the female are about half the size of normal wings and are quite without function. Such wings are clearly vestigial.

In many cave animals the eyes are vestigial. Thus in a number of cave fishes the eyeballs are small, spherical bodies hidden under the skin and are of no use whatever. The eyes of certain subterranean crayfishes, insects, and salamanders are similarly useless and hence vestigial.

Snakes are commonly regarded as legless reptiles. But in the python (Fig. 2) a small claw can be see on each side of the vent, and these claws are supported by bones within the body in such a way that they are clearly vestiges of hind legs.

Snakes are also peculiar in the structure of their lungs. Most air-inhabiting vertebrates have two lungs, one right and the other left. Many snakes have only the right lung, the left being represented only by a small protuberance.

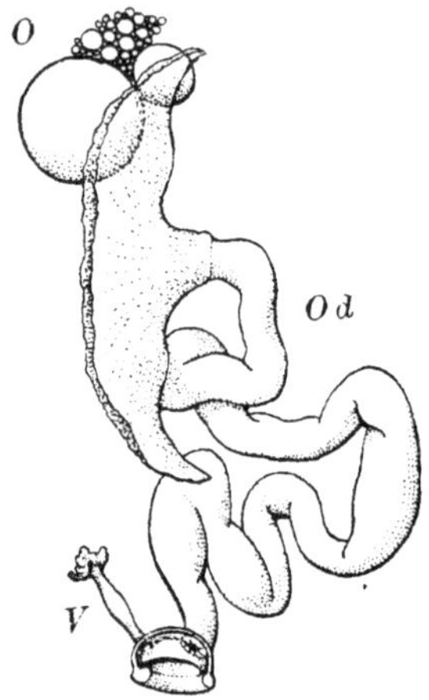

FIG. 3.—Reproductive organs of a female pigeon. *O*, Left ovary; *Od*, left oviduct; *V*, vestigial right oviduct. After Parker.

Vestigial organs are also well exemplified in birds. The ovaries and oviducts of most animals are evenly placed and equally developed on the two sides of the body, but in the birds these parts are functional only on the left side (Fig. 3). The ovary and the oviduct of the right side are abortive and quite useless, and hence vestigial.

The wings of certain birds are also vestigial. Wings used as organs of flight are among the most striking possessions of birds. Yet in the running birds, such as the ostrich and the cassowary, the wings are entirely useless for flight, for these birds are unable to rise off the ground. Relatively the feathers, bones, and muscles of the wings of running birds are very slight compared with those of the flying birds. In the New Zealand kiwi, the so-called apteryx, the wing is a diminutive member that is completely hidden under the feathers of the body, and hence the bird appears to be wingless. The wing of the kiwi is absolutely useless and is an excellent example of a vestigial organ.

Turning to the mammals we find many good examples of vestigial parts. Whales are mammals whose organization adapts them to a life in the sea. They are warm-blooded

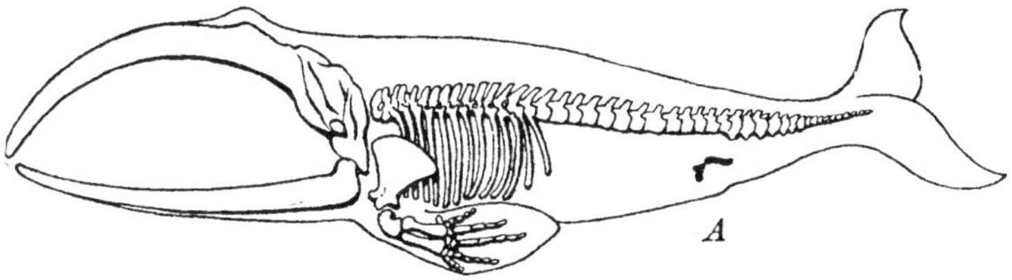

FIG. 4.—Side view of the skeleton of a whalebone whale, in which certain bones, shown in solid black at *A,* represent vestiges of hind limbs. After Weismann.

and are protected against changes of temperature by a covering of blubber instead of hair. They come to the surface of the water to breathe, and their blood system is so arranged as to allow them to store a large supply of purified blood, to be drawn upon during their submergence. Locomotion is accomplished chiefly by the enormous tail flukes, which spread out horizontally instead of vertically as do the tails of fishes. The flippers of the whale, which correspond to the forelegs of other animals, are used chiefly to guide these creatures through the water. Of hind limbs there is no external trace whatsoever, but when the interior of a whalebone whale is examined in the region where hind legs would be expected a group of isolated bones is found which correspond in part to the pelvis and in part to the legs of other mammals (Fig. 4). These bones are completely embedded in the substance of the whale and are apparently quite functionless. They are therefore good examples of vestigial organs.

The front and hind legs of the horse also contain vestigial parts. If the skeleton of the front leg of a horse is examined it will be found to be composed of many bones like those in the human arm. Buried in the flesh below the shoulder of the horse is a single bone corresponding to the one in our upper arm. Following this in the horse is a pair of bones duplicating the two bones of our forearm. Then comes, in both the horse and man, a small group of wrist bones. These are located in the horse at what is improperly called its knee. From the so-called knee in the horse to its hoof is a row of four stout bones, which correspond to the four bones in line with the middle finger of man. The first of these bones in the horse—that is, the one next

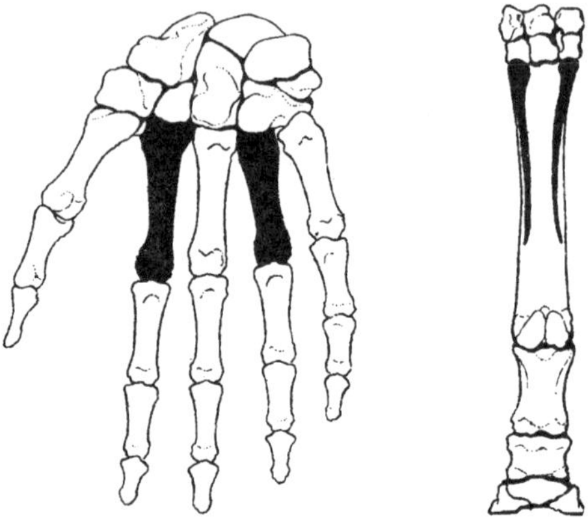

FIG. 5.—*A,* Palmar view of the skeleton of the hand of man, showing the wrist bones and the bones of the fingers; *B,* Similar view of a part of the front leg of a horse, showing wrist bones and the bones corresponding to the middle finger of man.

In the leg of the horse the vestigial splint bones are represented in solid black, and in the hand of man the corresponding bones are also shown in black.

the wrist bones—is the longest and corresponds to the long bone in the human hand that extends from the wrist through the region of the palm to the base of the middle finger. If this single long bone in the horse is examined it will be found to have on either side of it delicate splint bones, which lie in the flesh of this portion of the horse's legs but which support no parts in particular (Fig. 5). In position these two splint bones correspond to the long bones of our second and fourth digits, but they are not continued as such in the foot of the horse. The splint bones of the horse's foot are obviously remnants of other digits and are in the strictest sense of the word vestigial organs.

Turning to man we find him no exception to the rule that vestigial organs are abundantly present in the organization of animals. The external ear of the human being is a complicated fold of skin supported within by cartilage or gristle and occupying a fixed position on the side of the head. So far as hearing is concerned it is probably an organ of no great value. At least its occasional loss works no serious detriment to the hearing of its owner. But in listening for very faint noises we commonly extend the ear by holding a hand behind it, so that this organ probably serves somewhat as a collector of sound.

Notwithstanding the functional insignificance of the human external ear, this organ is provided with a rather remarkable group of muscles (Fig. 6). Extending from what may be called its root outward to the surface of the head are three considerable muscles, whose respective actions would be to bend the ear forward, upward, and backward. In addition to these muscles the surface of the ear proper has upon it six or more small muscles whose contractions would change slightly the form of the ear. All these muscles are well developed in the ears of certain lower animals, such as the horse and the dog, in which the three extrinsic muscles direct the ear as an ear-trumpet might be turned in relation to a source of sound, and the muscles of the ear proper change slightly the shape of that organ to adapt it better, perhaps, for the reception of a given sound. The human ear is incapable of these movements. The muscles, though present, are ordinarily functionless, though occasionally a person will be found who can move his ears slightly and in this way demonstrate a limited control over some of these muscles, but even the movement he produces is so slight as to have no advantage whatever for hearing and to be rather a *lusus naturae* than an act of physiological importance. In the horse and the dog the movements of the ears are of great value in discovering the direction of sound, but the muscles of the external ear of man perform no work comparable to that of the ears of lower animals. In man the muscles of the external ear are, in the strictest sense of the word, vestigial.

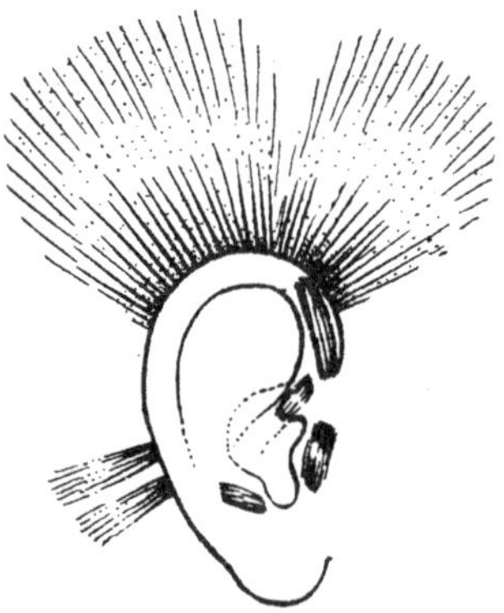

FIG. 6.—The human ear, showing the three extrinsic and four of the six intrinsic vestigial muscles.

Not only does the ear of man exhibit vestigial organs but a similar organ is found in the eye. Deep-seated in the nasal angle of the eye of man is a crescentic ridge of whitish tissue which, in consequence of its shape, is called the *plica semilunaris* or semilunar fold. It is not an organ that plays an important part in the action of the eye; in fact, it appears to be little more than a mechanical duplicature in the membrane in adjustment to the surface which it covers, and no one would suspect its meaning until he had examined the eyes of lower animals.

VESTIGIAL ORGANS

If we examine the angle of a cat's eye corresponding to the nasal angle in man we shall see there not a small fold of membrane but a veritable third eyelid. This additional lid may be made to pass under the two outer lids and over the whole exposed surface of the eyeball in a way to sweep this surface completely. It is provided with muscles by which this movement is quickly and easily made. This nictitating membrane, as it is called, is well developed in most mammals and serves to protect the eye. In man its representative is the insignificant fold already described, which, devoid of muscles and other appurtenances, is in its present condition a purely useless part, a vestigial organ. It is a noteworthy fact that, though the nictitating membrane is a functional organ in many mammals, it is reduced to a semilunar fold not only in man but in the monkeys and the apes.

Vestigial organs in man are found not only on the exterior of the body, as shown by the eye and the ear, but in its interior. In human beings the small intestine is not continuous with the large one end to end, but the small intestine enters the side of the large intestine, the natural termination of which is a pocket that projects backward and is known as the caecum. This pocket carries on its surface a small worm-like attachment, the so-called vermiform appendix, whose cavity opens directly into that of the caecum. In a way the vermiform appendix marks the real ending of the large intestine (Fig. 7). As is well known, the appendix readily becomes a center of intestinal disturbance and is the seat of the disease known as appendicitis, the usual surgical cure for which is to remove the appendix. Thousands of human beings have had the appendix removed and continue to live without experiencing any inconvenience. In fact, their position in life after the removal of this troublesome organ is so much more secure than it was before that it is now customary, as an incidental step in many abdominal operations, to remove the appendix on the assumption that a person is better off without it than with it. There is no reason to suppose that the appendix in man performs any function whatever, and since its removal is unattended by any subsequent inconvenience, it is commonly set down as a useless organ.

In other mammals than man the vermiform appendix or its equivalent presents a variety of conditions. It is not always easy in these lower forms to distinguish the exact limits between caecum and appendix, but in the rabbit, for instance, a large and highly complex caecum communicates freely and easily with an extended and apparently highly functional appendix. Without doubt these two parts are of great value in the digestive functions of this mammal, and the same may be said of them in many other animals. They are both greatly reduced, however, in the monkeys and in the anthropoid apes, where they are represented by a condition almost exactly like that in human beings. In consequence of the state of the vermiform appendix as seen in man it may be set down in the human species as a truly vestigial organ.

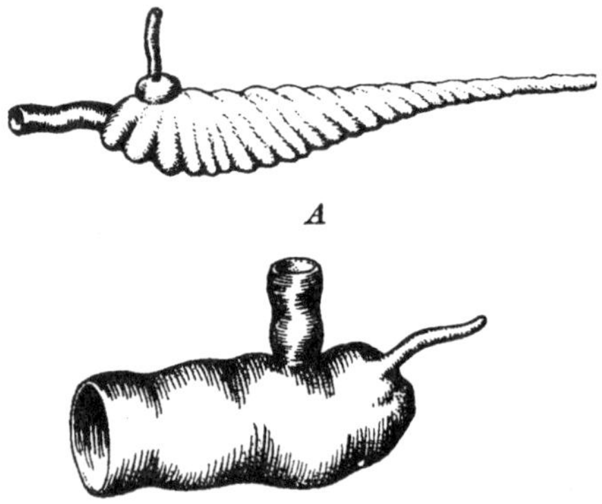

A

B

FIG. 7.—*A*, Cæcum and appendix of a rabbit, showing the small intestine entering from above and the large intestine emerging to the left; *B*, Similar parts in man, with the vestigial vermiform appendix to the right.

VESTIGIAL ORGANS

The so-called wisdom teeth of man under certain conditions partake of the nature of vestigial organs. As is well known, the milk dentition of a child contains in each half of each jaw two incisor or cutting teeth, one canine or eye tooth, and two molars or grinding teeth, making twenty milk teeth in all. When these teeth are shed a permanent tooth takes the place of each milk tooth and in addition three extra teeth appear in each half-jaw. These are the permanent molars, and their presence increases the permanent teeth to thirty-two. The last of these permanent molars at the back of each jaw is known as the wisdom tooth. Ordinarily the wisdom teeth are cut when the person is between twenty and twenty-five years of age, and they get their name from the belief that at that age the person has arrived at years of discernment. Not a few fail to cut these teeth or, in fact, even to form them at all. In such persons the number of teeth is four short of the usual total. Teeth that fail to cut the gums are of course useless and, in fact, like the vermiform appendix, they may be worse than useless, for such imperfect teeth may at times form centers of disturbance that call for surgical treatment.

This occasional reduction in the permanent dentition of man is in a way foreshadowed by what is seen in the monkeys. The new-world monkeys have a permanent dentition composed of a total of thirty-six teeth; the old-world forms, including the gorilla, chimpanzee, and other anthropoid apes, have four fewer permanent teeth and agree in this respect with man. Man appears to be going one step farther and to be reducing his dentition by dropping out another group of four teeth, the wisdom teeth, a step which, if finally taken, would place his permanent dental outfit at twenty-eight instead of thirty-two teeth. In man wisdom teeth that fail to cut the gums are, strictly speaking, vestigial organs, for under such conditions they are absolutely useless.

Not only is man the possessor of numerous vestigial organs in his adult state, but he also exhibits organs of this kind in his early stages of growth. Just within each nostril of the human embryo, or even of the new-born babe, is a small pore on the median wall of the nasal chamber (Fig. 8). This pore leads into a short, blind tube in the nasal wall. The pore and tube occupy exactly the position of Jacobson's organ of the lower vertebrates. This organ is an accessory organ of smell, which is well developed in many mammals and other vertebrates. In vertebrates other than man Jacobson's organ is provided with branches from the olfactory nerve, the nerve of smell, but in man this innervation is said to be lacking. As the organ disappears in man with the passing of childhood, and as it never shows signs of functional activity, it may be recorded as a vestigial organ of embryonic and early post-natal life.

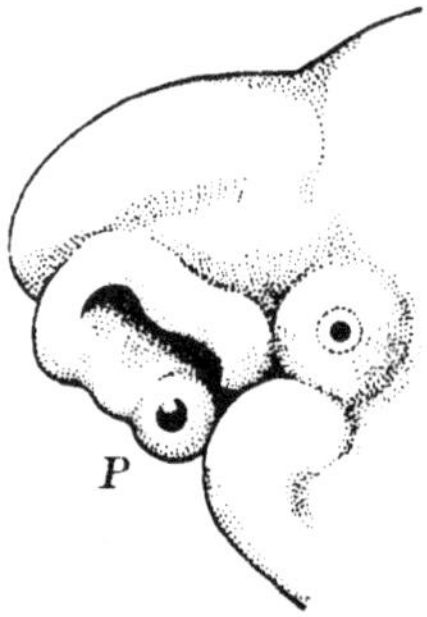

FIG. 8. — Side view of the face of a human embryo, showing the pore *P* of the vestigial organ of Jacobson at the entrance to the nasal cavity. After His.

Another vestigial feature, prenatal in time of occurrence, is seen in the lanugo of the human embryo, the covering of fine woolly hair found on the skin of the unborn human infant. This hairy covering is ordinarily shed before birth, and the separate hairs may often be identified in the amniotic liquor in which the embryo is immersed. This hairy covering is like that of the foetus of most mammals. At no time can it be of functional importance to the human embryo, and as it is ordinarily lost before birth it must be regarded as a vestigial organ, purely embryonic in its history. Its occasional retention after birth gives rise to the hairy men and women of the museums and the side-shows.

Man, both in his embryonic and his adult state, possesses an abundance of vestigial organs. In fact, students of this subject who have tabulated these parts have attributed to the human being almost a hundred such organs, and though some of these may on further investigation prove not to be true examples of vestigial parts, most of them certainly fall into this class, so that man may be said to be rich in organs of this kind.

The brief survey that has just been made shows that vestigial organs are widely distributed throughout the animal kingdom, and that they may be abundantly present in a given species, such as man. Their evolutionary significance has long been a matter of comment. If animals were specially created why should there be included in their bodies parts that are quite useless and often in fact positively detrimental to them? Why, for instance, should man possess a system of functionless muscles for his external ear, a useless hairy covering before birth, and a worse than useless vermiform appendix? No advocate of the theory of special creation has ever been able to give a satisfactory answer to these questions. To those who believe in special creation the presence of vestigial organs has proved a stumbling block that they have never been able to avoid. In fact, the occurrence of organs of this type has always been an insuperable obstacle to the acceptance of this view of the production of organic species.

From an evolutionist's standpoint, on the other hand, vestigial organs are precisely what should be expected. They are organs in process of disappearance. In the course of evolutionary change certain organs might naturally lose their usefulness and be replaced by others of a more appropriate type. As such organs gradually decay, so to speak, they should be expected to appear in much the way that vestigial organs do. Man possesses a system of functionless muscles connected with his external ear because these muscles were once useful to that organ in a pre-human ancestor who had occasion to move his ears as some modern animals still do. Man has a hairy covering before birth in consequence of his derivation from a stock of animals once fully covered with hair. His vermiform appendix is the remnant of an organ that was once a functional part of the digestive system of a remote ancestor.

Like other animals, man is not only a highly equipped and efficient organism with a most marvelous system of parts adapted to serve his bodily needs but he is also a repository of some of the most interesting and important relics of the past, relics whose significance can be truly understood only if they are viewed from the standpoint of the evolutionist. These relics are vestigial organs, and it is in this way and in this way only that such organs can be understood.

Organic evolution is not a principle that is open to direct and simple proof. Like the movement of the earth around the sun it can be demonstrated only indirectly. We do not even know that the earth is round by direct inspection. The shadow cast by the earth on the moon in an eclipse, the appearance of the ship as it rises over the horizon, and a number of other occurrences in nature are best explained on the assumption that the earth is round. The Copernican theory explains astronomic phenomena, it accounts adequately for all happenings in the skies. In a similar way organic nature, plants as well as animals, is full of happenings that call for some general explanation, and no principle

VESTIGIAL ORGANS

meets this call so surely as evolution does. Among these natural occurrences nothing is so difficult to understand, except from the evolutionary standpoint, as vestigial organs. These organs are really signs of the past; they afford as indisputable a proof of the correctness of the evolutionary view as can reasonably be expected.

GEORGE HOWARD PARKER

DARWIN FACTIONALISM

"Darwin himself would have turned his back on that theory (Evolution) if one single fact could have been produced in favor of the hypothesis of immutability, special creation, or supernatural agency. No such fact was forthcoming in his time, nor has any such fact been brought to light since."—Dorsey.

DORSEY

SOME DO - SOME DON'T

Many kinds of animals and plants exist to-day that show no records, and many kinds that do not exist to-day have left their records, in the rocks. If the rocks tell a true story, the story they tell is Evolution.—Editor.

FRANCES MASON

"Lunar years are the results of the appearance of the full moon each month. There are 12 Lunar years in our normal 12 month year. Noah lived 950 years. 950 divided by 12 equals 79 years and 11 months, or 950 months equals 79 years and 11 months, or 950 months."

BILL HUTH

EVOLUTION OF THE ORGANISM

ACCORDING to the theory of evolution as it is applied to zoölogy the fundamental likenesses or homologies among animals are the expressions of blood relationship. We believe, for example, that all the various species which belong to the cat tribe—such as the domestic cat, the leopard, the jaguar, the puma, the lynx, the lion, and the tiger—are the descendants of a primeval species of cat, just as the dog, the fox, the jackal, and the wolf are the descendants of a primeval species of dog. Furthermore, we believe that the primeval dog and the primeval cat were distant cousins, and that millions of years ago both were represented by one species of primitive carnivore, from which both have been derived.

Now the human body resembles the body of a higher ape just as the body of the cat resembles that of the dog; and if our ideas as to the relationship between cats and dogs are sound it must follow that apes and men have been gradually developed out of one and the same ancestral species. But we assume that evolution proceeds very slowly—that great changes require millions of years—so that direct evidence of it is impossible to obtain. To get such evidence, indeed, we should require the notes of some immortal angelic observer who had watched the process of evolution for ages and had left us a record of his observations! All our evidence must therefore be indirect and circumstantial, and it may be divided into three main groups—(1) evidence from the systematic relations of allied species, (2) evidence from fossils (palaeontology), and (3) evidence from the development of the individual (embryology). The evidence comprised under the first two heads is presented elsewhere in this book; it is my task to present the evidence or argument from embryology.

The embryological argument rests on a daring hypothesis, which was first clearly expounded by Haeckel in what he termed *the fundamental law of biogenetics,* which he enunciated as follows: Every animal, in its growth from the egg to maturity, *recapitulates* the history of the race. Therefore, when we find in the embryo, during its growth, stages that resemble animals lower in the scale of life, this resemblance is regarded as evidence that the animal whose embryo we are studying was derived from ancestors that resembled those stages.

When we take a broad survey of the animal kingdom we find that it can be divided into great primary divisions termed phyla. The bodies of all the members of a phylum are constructed on the same plan. A good example of a phylum is provided by the Arthropoda, the jointed animals. This group comprises all the insects, spiders, mites, scorpions, centipedes, lobsters, crabs, shrimps, barnacles, and water-fleas; in fact, it includes three-fourths of the species of animals now living on the globe. All these animals have, outside their flesh, a hard, shelly skeleton, which is divided into joints by zones of thinner shell, as otherwise the animal could not bend itself or move. All likewise possess several pairs of limbs, which, like the body, are encased in similar armour and similarly provided with joints or hinges. Each phylum is divided into classes, the members of which agree in their fundamental plan. Thus the Arthropoda (if we exclude certain smaller groups) are divided into three classes (1) the Insecta, in which there is a pair of feelers in front of the mouth and in which the body consists of three parts, the head, the thorax, and the abdomen; (2) the Arachnida (the scorpions, spiders, and mites), in which there are no feelers, but only a little pair of pincers in front of the mouth, and in which none of the legs are changed to jaws; and (3) the Crustacea, in which there are two pairs of feelers in front of the mouth and several jaws. These three broad classes are divided into orders, the orders into tribes, the tribes into families, the families into genera, and the genera into species. The Malacostraca form an order of Crustacea in which the body is divided into just twenty segments, and this order is classified into two main divisions, the Macrura, or the long-tails (the shrimps, prawns, and lobsters), in which the abdomen is long, stretched out, and the Brachyura, or the short-tails (the crabs), in which the abdomen is reduced to a useless vestige and carried under the rest of the body. But there is an intermediate tribe, the Anomura, which includes the hermit crabs. A member of this tribe has an abdomen of moderate length, and many of them have the extraordinary habit of thrusting it into an empty snail shell, in consequence of which habit it has become curved and asymmetrical. Now no one who has studied animals of this type doubts that they are modifications of the type having a straight, symmetrical abdomen. If we trace the course of the development of these twisted-bodied crabs we find that when they are young they swim in the sea like little shrimps and that during this stage of their lives the abdomen is perfectly straight; so that the life history confirms the conclusion to which we were led by the study of the adult form —that the twisted body is a modification of a straight one. These crabs with the aborted abdomen must have been developed from animals—such as the shrimps and lobsters— in which the abdomen is long and normal; and so we find that in their young stages they are very much like little lobsters. If these crabs were specially created why should they begin life like lobsters?

In fact, we may lay down the general law that *any member of an order or tribe that differs in structure from the other members will show in its young stages a close agreement in structure with the type prevailing in that order or tribe.*

A splendid example of this law is provided by the Indian cat-fish *Clarias.* The cat-fishes are widely distributed over North America, Europe, and Asia, and their general appearance is familiar to most persons. With one or two exceptions all are river fish and are quite devoid of scales, having naked dark-brown or black skins, flat heads, and broad mouths. From both lips and from the corners of the mouth protrude long, slender rods called barbels, with which these fish probe the mud for the worms and other small animals on which they live. It is the fanciful comparison of these barbels to the whiskers of a cat that suggested the name cat-fish. Now one of the genera of Indian cat-fish called *Clarias* has learned to emerge from the water and breathe air, and in order to do this it has developed above its gills two curious tree-like organs, which are richly supplied with blood vessels. As Figure 1 shows, however, this fish is essentially like other cat-fish when it is young, but as it grows older two little buds appear above the gills on each side, which develop into the tree-like organs of the adult. Can anyone seriously doubt that *Clarias* has been developed from an ordinary cat-fish? And if we admit this must we not also admit that, here, at least, the young animal recapitulates the past history of the race?

In recent years the law of recapitulation has been proved by experimental evidence. It has been found that the black and yellow salamander of Europe (*Salamandra maculosa*) is capable of slowly altering its colour as it grows up; it makes its colour harmonize with that of its surroundings. If these young salamanders are confined in yellow boxes the yellow spots on their skins enlarge in size as they grow to maturity. If they are confined in black boxes the yellow spots diminish in size and many of them disappear. These changes are in some degree passed on to the offspring, for if two "yellowed" salamanders are mated together they produce young that are much yellower than their parents were at the corresponding stage of development, and if these young continue to live amid yellow surroundings they become almost entirely yellow when they are fully mature. Now if the offspring of "yellowed" salamanders are reared in black boxes *they steadily become yellower for the first year of their lives,* thus recapitulating the experience through

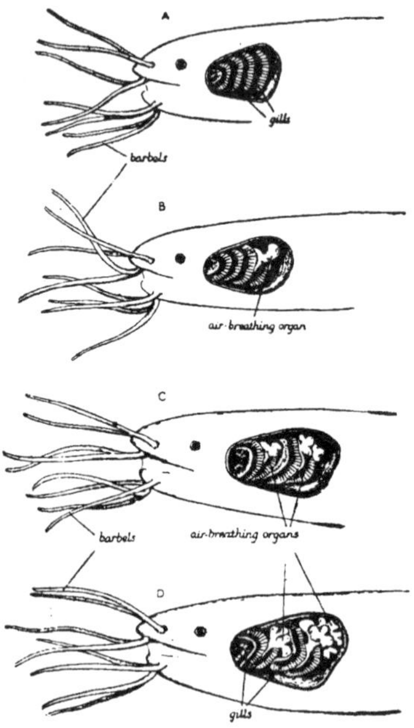

FIG. 1.—Stages in the development of the Indian cat-fish *Clarius* from an ordinary cat-fish.

which their parents passed; after that, and then only, does the effect of the black environment begin to tell on them, for the yellow spots begin to diminish in size.[1]

The young forms termed larvae, which by their structure and habits repeat ancestral conditions, live freely in the world and earn their own living. The most familiar example of a larva is the tadpole of the frog, which, by its gills and tail recalls the fish-like ancestors of the frogs. There is another type of young animal which is known as an embryo (Greek εν (εμ), in, βρύειν, grow). This type, during its development, is sheltered and fed either within an egg shell or in the womb of the mother. A good example of this type is the chicken, of which the greater part of the development is completed within the egg shell. The young form in this type derives its food from the yolk in the egg, which it slowly digests as it grows, and from the "white," or albumen. Another variety of embryo is sheltered within the womb of the mother and obtains all its nourishment from the maternal blood. The embryos of all the higher warm-blooded mammals, such as those of dogs, horses, and cattle, as well as the human embryo, are of this kind.

Now let us consider how the embryonic and the larval types of development are related to one another. Was the original form of development embryonic or larval?

When we closely examine the life histories of animals we discover that there is an embryonic and a larval phase in all, though these phases are of extremely different lengths in different animals. No animal deposits naked eggs; an egg-shell, though it may be thin and elastic, is always formed, and the egg always includes some yolk, so that there is always a period during which the young animal develops as

[1] All biologists are not agreed as to the sufficiency of the evidence for the inheritance of acquired characters.

FRANCES MASON

an embryo before it bursts the egg-shell and begins as a larva to earn its own living. Furthermore, an animal is not hatched in exactly the form of the adult; it attains this form only after a period of growth, and during this period it may properly be termed a larva.

The salamanders show us clearly that the embryonic phase is a secondary modification of the larval phase—in a word, that the embryo is a larva which is provided with shelter and food.

There are two species of European salamander, the black and yellow salamander (*Salamandra maculosa*), commonly known as the fire salamander, which inhabits the plains and lower reaches of the valleys, and the black salamander (*Salamandra atra*), which is found in Alpine pastures. Both species bring forth the young alive, but the young of the fire salamander are provided with long, feathery gills and gill slits and pass the first six months of their life in water, whereas the young of the black salamander are devoid of gills and gill slits at birth and are ready to take up the parental habit of life on land.

If now we open the womb of a pregnant black salamander we shall find in it a number of embryos with long, feathery gills, and if these embryos are thrown into water some of them will survive and develop like the young of the fire salamander. The gills are an adaptation to life in water, and when we find them in an embryo we may be sure that the embryo was once a larva and that the larval phase is therefore the primary one.

Similar examples could be adduced from almost every group of animals. We have already alluded to the well-known tadpole larvae of frogs and toads. There is, to mention one more animal, a small West Indian tree frog (*Hylodes*) which produces a few large eggs from which emerge not tadpoles but little frogs, which resemble their parents. If, however, we dissect off the membranes of developing eggs we find within them tadpoles, complete, with their characteristic tails.

The earliest stages in development are the most delicate and vulnerable, and it is these which first become embryonic; the latest stages in development, which represent comparatively recent ancestral history, are always larval. In human development, as we all know, the baby is an embryo for nine months before birth, and after it is born the child may be justly termed a larva until the beginning of puberty. The mental powers are not fully developed until the child reaches the age of about fifteen years.

During the period of its life within the womb the human embryo develops a large organ like a sucker, which is closely pressed against the wall of the womb and which enables the tiny baby to suck nourishment from its mother's blood. This sucker, which is called the placenta, is developed from the belly of the embryo, which is thereby distorted out of shape. Now no one imagines that some ancestor of man went about through life with a placenta protruding from its under surface; the placenta is a secondary outgrowth to enable the embryo to live in the womb. Such "secondary" changes are known as falsifications of development; they may be likened to interpolations made by some later writer in an ancient historical document. But during the time that the embryo carries this extraordinary appendage, protruding from its under surface, its upper surface passes through a most interesting series of changes. Its mouth at first resembles that of a shark, and the nostrils, as in the shark, are connected with the edges of the mouth by grooves. Then the head grows to be like that of the tadpole, and, just as in the young tadpole, this head is divided from the body by a narrow neck, which has nothing to do with the neck of the young child. Along the sides of the neck there are two series of gill slits, and, just as in the tadpole, these become covered by flaps of skin that grow back from the head and join the trunk. The neck indentation is thus obliterated, and the head passes without a break into the trunk, just as it does in the older tadpole. The blood vessels at the sides of the gill clefts resemble exactly those of the tadpole. There are four of them on each side, and, with accurate imitation of the tadpole, the third on each side drops out. The salamander retains the four throughout life, but its near cousin, the newt, drops out the third, as does the frog. Thus the story of man's development from a water animal and his gradual closing up of his gill clefts is accurately repeated in the womb, and the distortion of this story by the development of the placenta is easily recognised. We find the same history if we study the development of the young lizard within its mother; but here no placenta is developed, and the egg is afterward laid, but development has begun long before that. So by comparing the life histories of different animals belonging to the same phylum we can separate the secondary accretions from the original story and thus recover the true ancestral history.

To return to human development: As this proceeds the limbs grow out and the embryo comes to resemble an ordinary four-footed animal, but the fingers and toes are at first webbed like those of a frog. At this stage there is a well-developed tail, and later there is a complete covering of hair, resembling the hairy skin of an ape. At birth the big toe is widely separated from the other toes, just as is the big toe of an ape, and the legs curve inward at the ankles, so that when the child is held upright only the outer edge of the sole rests on the ground. This arrangement of the legs is identical with that seen in the leg of an ape; it is an adaptation that makes it easy to press the soles of the feet against opposite sides of a branch in climbing.

The repetition of ancestral phases in life histories has been compared to memory. If we learn tennis when we are young and have no time to practice it until middle life we shall nevertheless find that when we resume the game a part of our skill comes back to us, that, in a word, we remember it. So when salamanders have grown up in a yellow box and have become yellowed in consequence, their young "remember" the parental experience and strive to turn yellow in spite of being enclosed in black boxes. The tadpole represents a memory of the time when the ancestors of the frog lived like fish. But all development does not reflect purely ancestral experience. Just as memories become blurred and sometimes confused with the lapse of time, so secondary modifying factors tend to render the ancestral record of development illegible. The organs of a larva are used, broadly speaking, in much the same way as the ancestors used them, but the corresponding organs of an embryo are not used, and so these tend to be imperfectly developed and to degenerate into mere sketches of the ancestral organs. Larval life, which represents ancestral habits, may become excessively dangerous, and the larva, in self defence, may be driven to adopt another mode of life, as when the young of the May-fly learn to live in water and develop secondary gills. But these secondary modifications are usually peculiar in each kind of animal, and the fundamental ancestral record can be deciphered only by comparing a large number of life histories, just as the historian extracts the truth from ancient documents by stressing the points in which they agree and discounting those in which they differ.

The most direct proof of evolution is certainly afforded

by fossils. When we study the fossils preserved in a great mass of strata, overlying one another like the pages in a gigantic book, we find some of these fossil animals gradually changing as we pass up the series. In the Devonian strata we find numerous remains of fish with fins consisting of a central axis beset with two rows of branches like a feather. As we reach the base of the Carboniferous strata these give place to newts with five-fingered hands and feet. An eminent palaeontologist (Prof. D. M. S. Watson, F.R.S.) who has studied this series of fossils, arrived at the conclusion that the fourth finger on the hand represented the axis of the feather, that the fifth finger was the sole remaining branch on one side, and that the first three fingers were the branches on the other side. A German embryologist, who was totally ignorant of Dr. Watson's conclusion, studied the early development of the wing of the common fowl. Here, in the youngest stage, five columns of condensed tissue, representing the five fingers, can be made out, though in the adult bird only two and the trace of the thumb remain. From a study of the growth of these rudimentary fingers the German arrived at the same conclusion that was reached by Dr. Watson from his study of the fossils, namely, that the fourth finger represents the axis of the fin. Thus, when the opportunity is afforded, the conclusions drawn from embryology are confirmed by palaeontological evidence. Other examples of the same confirmation could be given, but their citation would involve detailed descriptions of anatomy. As, however, these confirmations are increased in number, our confidence in the truth of the embryological record grows; and this confidence is of the utmost importance to us for tracing the history of evolution, because this ancestral record, repeated in embryonic development, gives us the only means that we possess of tracing the history of life back to its beginnings, or rather of completing the imperfect story told by fossils.

The earliest stages in this history in the human embryo are blurred out of all recognition, but if we examine the earlier history of the tadpole or, still better, that of the newt, a primitive amphibian, we can get a general idea of the history of life from the beginning. The egg, as we have seen, is a single cell in its essential structure, the same as a whole group of single-celled animals termed the Protozoa. This cell divides into many cells, which cohere together and form a little hollow ball called the blastula. Similar little balls, in which the cells have acquired green colouring matter, roll about in the waters of our ditches. The blastula becomes converted into a hollow cup called the gastrula by the pushing in of the cells at one end. The opening of the cup is called the blastopore. This opening is retained throughout life as the anus or vent of the animal, and above the anus the tail grows out. The outer layer of the cup forms the skin; the brain and the nervous system are at first mere thickenings in this layer. The inner layer forms the lining of the stomach; the rudimentary backbone is only a ridge or folding of this inner skin along its upper surface. The mouth is formed as a new opening in front; the gill slits are clefts at the sides of the throat; the eyes grow out as buds on the brain; the nose and ears are pits in the skin, and behold! we have before us no longer a swimming cup but the beginning of a tadpole, which is really a very primitive type of fish.

A study of the development of jointed animals (the Arthropoda), of the Mollusca (clams, oysters, and snails), and of the Echinodermata (starfish and sea-urchins) leads us to similar startling and fascinating results. We find that worms, arthropods, and mollusks arose from a common ancestor and diverged from one another owing to their adoption of different habits; and, strangest of all, that backboned animals, the class to which we ourselves belong, diverged from the same root as the starfish, sea-urchins, and sea cucumbers, which constitute the class Echinodermata.

Comparative anatomy and systematic zoölogy take us only a little way, for we have no reason to assume that the ancestral forms of animals have persisted unchanged to the present day. The evidence from fossils is best, but fossils preserve only the hard parts, and the earliest fossils thus far found are already far advanced in evolution. But every animal begins its development in the egg, which is a single cell, comparable in structure to the lowest forms of life known to us, and as it grows to the adult form it sketches in broad outlines the whole story of its evolution.

ERNEST WILLIAM MCBRIDE

BROTHERS OF TRUTH

The date, February 12, 1809. The place, Shrewsbury, England. Yes My Brothers, for not only are We celebrating the birth date of Abraham Lincoln on this day, but also in this exact same year, and on the exact same day, another great emancipator was born to mankind . . . . Charles Darwin, the man who freed mankind from the ignorance of "Adam."

Remember this Lincoln said, *"all are created equal,"* Darwin proved it . . . .

BILL HUTH

EMBRYOLOGY AND EVOLUTION

In early times and among primitive peoples all phenomena were regarded as supernatural. The rising and setting of the sun, the sweet influence of the Pleiades, the coming and going of the winds, storms, lightning, thunder—all the phenomena of life, birth, and death—were supposed to be directly controlled by gods or spirits. In the course of centuries many such events were seen to be natural—that is, lawful or orderly—and were more or less understood, so that gradually the supernatural withdrew to the misty mountain tops of origins. During the last two or three centuries enlightened people everywhere have come to realize that ordinary phenomena occur in accordance with natural laws. But in the matter of beginnings and origins the opinion is still widely held that they do not happen in accordance with nature, but only in response to supernatural action.

The Nature of Development

Even such a constantly recurring phenomenon as the origin of the individual human being was by many regarded as a supernatural phenomenon until a little more than a hundred years ago; and even today many intelligent people believe that the mind and soul of every person is supernaturally created, though few, if any, would go so far as to maintain that the body is supernaturally created, as some of the "preformationists" did in the eighteenth century. This old doctrine of preformation, or "evolutio," as it was called, maintained that the fully-formed but minute organism was encased within the egg or sperm, and the microscopists of the day, with poor instruments but good imagination, thought they could see the "homunculus," or little man, neatly packed away within the human germ cell. Within this homunculus in turn it was held that there must be another generation of germ cells, each containing its homunculus, and so on *ad infinitum.* Thus arose the doctrine of infinite encasement, or "box in box," and as the schoolmen of the Middle Ages discussed how many angels might stand on the point of a needle, so their successors of the eighteenth century discussed how many fully formed but infinitely minute generations might have been contained in the ovary of Mother Eve. These speculations reached their culmination in the works of Charles Bonnet (1748-1773), the distinguished natural philosopher of Geneva, in which he denied all new formation, all development or generation, and held that in the original creation of the progenitor of each species God created at one stroke all the individuals that would ever come from that progenitor. Thus every individual in the world was supernaturally created.

The actual study of the development of eggs forever put an end to such speculations. Caspar Frederick Wolff (1759) demonstrated that fully formed but minute organisms are not contained in germ cells; that development is not a mere unfolding of that which is already infolded, but that it consists, from inception to maturity, in an increase of complexity; and though he over-emphasized the simplicity of the germ, no one now questions that individual development everywhere consists of progress from a relatively simple to a relatively complex form. Development is not the unfolding of an infolded organism; it is the formation of new structures and functions by combinations and transformations of the relatively simple structures and functions of the germ cells. If anything in the world is natural, this process of development from an egg is natural; but nothing in the world is more wonderful. And yet this process of development is according to nature—a natural development and not a supernatural creation. Religious faith has been able to survive the knowledge of the fact that every human being in the world has come into existence by a process of development. Why should it be supposed that the recognition of an equally natural development of groups of individuals or species would be destructive of religion?

Ontogeny, or the origin of individuals, and phylogeny, or the origin of races, are two aspects of one and the same thing, namely, organic development. There is a remarkable parallelism between the two, and in particular the factors or causes of development are essentially the same in both. Just as the earth rotates on its axis and revolves in its orbit and the whole solar system moves through space in accordance with the law of gravity, so organisms undergo development as embryos, as species, and as larger groups according to the law of organic evolution.

Although the word evolution is now used to specify the development of species and of larger groups of organisms, it was once applied also to the development of individuals, and it could still be so applied, for *evolution means only the transformation of an earlier into a later stage according to natural laws.* In short, evolution is trans-formation rather than new-formation, natural development rather than supernatural creation, whether we are considering the origin of individuals or of species or larger groups.

VOICE OF CREATION

The Course of Development

Let us consider in brief outline some of the main facts of individual development. In practically all animals and plants development begins with the fertilization of an egg. William Harvey (1651), the discoverer of the circulation of the blood, expressed this fact in his famous dictum, "Omne vivum ex ovo." The egg is a cell with the structures and functions that are characteristic of cells in general; that is, it contains protoplasm, "the physical basis of life," and this protoplasm is differentiated into a nucleus and a cell-body, and each of these contains other smaller units, many of which differ one from another. But these units are not adult parts in miniature, as the "preformationists" supposed; they are merely the elements out of which adult parts are built; they are like the letters of the alphabet out of which are built words and books and literatures.

In all animals the process of fertilization is practically the same, consisting essentially in the union of the nuclei of egg and spermatozoon. In almost all multicellular animals the egg at or near the time of fertilization gives off two minute cells, the polar bodies, which are rudimentary eggs and take no part in development. The fertilized egg undergoes repeated divisions or *cleavages,* forming a mass of cells, usually a hollow sphere, which is called a *blastula,* and this in turn becomes a *gastrula* by the formation in it of a gastric cavity. So far all animals, from sponges to men, travel the same road, although in every group there are minor peculiarities; they travel the same road, but they do not all follow in exactly the same tracks. The germ cells, the cleavage, the blastula, and the gastrula show characteristic differences in different phyla or classes, but their resemblances are more significant than their differences, and within the same phylum these resemblances are very striking. For example, in groups as distinct as flat-worms, annelids, and mollusks corresponding cleavage cells give rise to similar organs, and the larvae of these forms are very much alike, though the adult forms are very different.

In all vertebrates, from Amphioxus to man, there are many fundamental resemblances which distinguish this one phylum from all others. The eggs of vertebrates vary greatly in size, ranging from the microscopic eggs of mammals to the enormous ones of birds, and they differ also in the conditions under which they develop. Some are set free as naked cells to begin their lives independent of their parents; others are enclosed in protective membranes and shells and are further guarded and incubated by the mother or father; and still others undergo their development within the body of the mother. Associated with these varying conditions of development are many differences in the size of eggs, the rate of development, the methods of nutrition and other features. But the striking fact remains that in all vertebrates, from the lowest to the highest, one finds fundamental agreement in the position and structure of the principal organs of the body, and these organs arise from corresponding parts of the egg or embryo in essentially the same manner in all. Thus the brain and the nervous system come from a plate of superficial cells on the dorsal surface of the gastrula—a plate that rolls up at the sides to form a groove and then a tube. The front end of this tube enlarges to form the brain; the hinder part forms the spinal cord. The backbone, which gives the name to the phylum (vertebrate or chordate) appears as a row of cells, the notochord, above the alimentary canal and below the nerve plate. This row of cells is the basis of the backbone; it becomes surrounded by cartilage, is then changed to bone and is segmented into the vertebrae of the vertebral column. In all vertebrates the right and the left side of the pharynx—the cavity behind the mouth and nose —are pierced by a series of gill slits, and between these slits run arterial arches, which in fishes and water-breathing amphibians bring blood to the gills to be aerated; but in all air-breathing vertebrates true gills are lacking, though the arterial arches are still present in the embryo, and they undergo characteristic modifications in different classes of vertebrates. This very brief and general statement of the course of embryonic development applies to all vertebrates, man included.

In all animals development begins with the fertilized egg, which contains none of the structures of the developed animal, though every egg does contain specific kinds of protoplasm, which differ not only in different phyla and species but even in different individuals of the same species. These different kinds of protoplasm in the germ cells constitute the material basis of inherited differences in mature organisms, and the more unlike animals are in their adult state the more unlike is this germinal protoplasm of the egg cells from which they develop. In all organisms the egg contains hereditary units by whose combinations and transformations the characters of the cleavage, blastula, gastrula, embryo, and adult are formed. Development consists in such combinations and transformations, and just as the chemical combination of hydrogen and oxygen to form water gives rise to a substance with new qualities which were not present in the elements that entered into the combination, so new structures and functions develop out of germ cells—structures and functions that were not present as such in the cells. This is "creative synthesis," or "creative evolution," and it is a phenomenon that is found in all nature.

The development of functions goes hand in hand with the

development of structures; indeed, function and structure are merely different aspects of life. All the general functions of living things are present in germ cells. In its growth every egg or sperm cell takes in nourishment, which it transforms into its own protoplasm; it divides at intervals and thus reproduces; it is sensitive, or capable of responding to stimuli. These are the fundamental functions of all living things, and every germ cell has them, but as development advances each of these functions becomes more specialized and more perfect. Nutrition, reproduction, sensation, which are all present in the egg cell, become, in the course of development, localized in cells specialized for each of these functions. But just as in the development of structures new parts, which were not present in the germ, appear by a process of "creative synthesis," so new functions appear in the course of development, which are not merely sorted out of the general functions that were present at the beginning but are created by the combination and interaction of parts and functions already present. In this way the highest and most marvellous functions develop out of egg cells—even the special senses, instincts, and higher psychical faculties of animals and man. All are products of development or evolution—that is, they have come as a result of new combinations and transformations of the functions present in germ cells. Every step in this process is natural; yet that a complex animal, even a man, with all his godlike faculties, can develop out of a germ cell is surely the climax of all wonders!

The Recapitulation Theory

At the beginning of the nineteenth century the belief was general that higher animals pass through stages in their development that correspond to the adult condition of lower animals. In 1828 von Baer, "the father of comparative embryology," recognized that this view was not wholly correct, and he modified it as follows: "An embryo never resembles an adult animal and is only to be compared with the embryos of other animals. The more different two animal forms are in their end stages the farther back in their development must one go in order to find agreement between them." This has often been called "von Baer's Law." Louis Agassiz, in his famous *Essay on Classification* (1858), pointed out the fact that there is a parallelism between embryology, palaeontology, and classification in that the stages that an animal passes through in its development from the egg resemble certain animal forms that have appeared in the past history of the earth and also certain lower forms now living.

The full significance of this parallelism was not appreciated until the revival of the doctrine of evolution under Darwin. In the fourteenth chapter of the *Origin of Species* Darwin discusses this parallelism and the significance of the homologies of embryos, and he closes his discussion of embryology with these carefully guarded words: "Embryology rises greatly in interest when we look at the embryo as a picture, more or less obscured, of the progenitor, either in its adult or larval state, of all the members of the same great class."

It was Ernst Haeckel, in his *Generelle Morphologie* (1866) and in many later books, who announced that "Ontogeny is a short recapitulation of Phylogeny"—that is, the successive embryonic stages in the development of an animal correspond to the successive adult stages of the phylum to which it belongs. This is Haeckel's Fundamental Law of Biogeny ("Biogenetisches Grundgesetzt"), which is more frequently called the theory of embryonic recapitulation.

EMBRYOLOGY AND EVOLUTION

Inasmuch as many phenomena of development are mere adaptations to the conditions of embryonic or larval life and could never have been present in adult animals, Haeckel separated such characters, which he called "cœnogenetic," from the truly ancestral ones, which he called "palingenetic." Unfortunately there was no certain method of always distinuishing these two types of embryonic characters, but in spite of this difficulty embryology was supposed to afford a short and easy method of determining the ancestral history of every group. Since every animal in its development from the egg to the adult condition was believed to climb its own ancestral tree, one can imagine the feverish zeal with which the study of embryology was pursued. Here was a method which promised to reveal more important secrets of the past than would the unearthing of all the buried monuments of antiquity—in fact nothing less than a complete genealogical tree of all the diversified forms of life which inhabit the earth. It promised to reveal not only the animal ancestry of man and the line of his descent but also the method of origin of his mental, social, and ethical faculties.

Unfortunately there was no certain criterion by which the palingenetic or ancestral features of development could be distinguished from the cœnogenetic or recently acquired ones, and what one embryologist regarded as ancestral another might consider a recent addition. Furthermore, when there were no living or fossil animals resembling certain embryological forms the fancy was given free rein to invent hypothetical ancestors corresponding to such forms.

As a result of such speculations multitudes of phylogenetic trees sprang up in the thin soil of embryological fact and developed a capacity of branching and producing hypothetical ancestors which was in inverse proportion to their hold on solid ground. For a time embryology was studied chiefly to learn the course of past evolution, but owing to the highly speculative character of such studies and to the differences of opinion as to what were original (palingenetic) and what were acquired (cœnogenetic) characters, there gradually arose a widespread skepticism concerning the value of embryology for this purpose. Gegenbaur, in 1889, voiced the growing opinion among zoölogists in these words: "If we are compelled to admit that cœnogenetic characters are intermingled with palingenetic, then we cannot regard ontogeny as a pure source of evidence regarding phyletic relationships. Ontogeny accordingly becomes a field in which an active imagination may have full scope for its dangerous play, but in which positive results are by no means everywhere to be attained. To attain such results the palingenetic and the cœnogenetic phenomena must be sifted apart, an operation that requires more than one critical *granum salis.*" Since the time this was written there have been many other less moderate utterances to the same effect, some even declaring that there is no evidence that ontogeny ever recapitulates phylogeny and that Haeckel's "biogenetic law" has no foundation in fact.

But after all, these criticisms of certain details of the recapitulation theory have not destroyed the general and fundamental truth of that theory—namely, that many features of individual development repeat ancestral features. There are many remarkable and undoubted instances in which ontogeny repeats phylogeny and in which the relationships of organisms can be determined only by their embryological history. The most severe critics of Haeckel's "biogenetic law" do not deny this; their criticisms apply to details rather than to foundation principles.

It is certainly no mere accident that practically all animals being their individual existence as fertilized eggs; that before or during fertilization all eggs produce two polar bodies, or rudimentary eggs, which undergo no further development; that in all animals fertilization occurs in essentially the same way; that all eggs undergo a series of divisions or cleavages which lead to the formation of a hollow sphere, the blastula, and that the blastula gets a digestive cavity and becomes a two-layered embryo, the gastrula, comparable to a simple sponge or hydroid. It is certainly no accident that the cleavage of the egg in types as distinct as flatworms, annelids, and mollusks is almost cell for cell the same, and that, where differences exist, rudimentary cells have been found that correspond to well-developed cells in other forms. It is no accident that the eggs of all chordates, for example, have the same type of organization, that all develop a notochord and nervous system and sense organs and gill slits and excretory organs and hundreds of other structures in essentially the same way. These fundamental resemblances, or *homologies,* as they are technically called, call for some explanation, and the only natural explanation that has ever been proposed is evolution.

Some of these homologies between different embryos and between these and adult forms of lower animals are worthy of more detailed mention. The earliest stages of ontogeny are in many respects like the lowest living organisms. The egg carries us back to the protozoan; cleavage recalls the protozoan colony; the blastula suggests volvox-like forms; the radial gastrula suggests the sponge or hydroid forms; and the bilateral gastrula suggests polyclad-like forms.

From the gastrula stage onward different phyla usually follow different paths, but all the members of each phylum show many fundamental resemblances in embryonic and larval stages, though they may differ notably in adult structure. For example, crustaceans that differ widely in adult form have larvae that are strikingly alike. Zoölogists had long classified barnacles as mollusks, until a study of their larvae showed conclusively that they were crustaceans. Many parasitic crustaceans are mere sacs of eggs or spermatozoa in the adult stage, but they have typical crustacean larvae. Ascidians, or sea-squirts, were classified as mollusks until Kowalevsky showed that their larvae are little tadpoles with notochord, dorsal nerve-tube, and gill-slits like any typical vertebrate. These and many other examples show that natural affinities or ancestral relationships are often shown in embryos and larvae long after they have been obscured or lost in adult stages.

All vertebrates, from fishes to mammals, pass through a fish-like stage in their development in which they have (1) gill slits, (2) five or six pairs of aortic arches, (3) a simple tubular heart with one auricle and one ventricle, (4) a notochord (the basis of the backbone in higher vertebrates), (5) a primitive type of kidney (the pronephros), etc. These organs persist throughout life in the lowest fishes, but they undergo many changes in higher forms.

Entirely similar conditions are found in the development of the brain and central nervous system; the eye and ear; the limbs and muscular system; the digestive, respiratory and reproductive systems. In fact, nearly all the important organs and systems of higher vertebrates pass through stages in their development which in lower vertebrates remain permanently.

It is true that some of these embryonic reminiscences of lower forms have been modified and greatly abbreviated, but they are nevertheless all there and are recognizable. Among higher forms there are many adaptations to peculiar conditions of embryonic or larval life which have no counterpart in lower forms; such are the embryonic membranes of higher vertebrates, modifications due to the presence of much yolk, and extraordinary modifications of parasitic larvae. Such modifications indicate that evolutionary changes may occur in embryonic as well as in adult stages—in fact, that they may affect any part of the life history. But this does not disprove the thesis that in general "ontogeny recapitulates phylogeny." The fundamental resemblances or homologies between embryos, larvae, and adults which have been cited above are just as genuine homologies as those between adult structures, and the only natural explanation that has ever been found for such homologies is inheritance from common ancestors.

The origin of the individual (ontogeny) is not only a key to the conditions and causes of the origin of the race (phylogeny) but is also an actual evolution—that is, an origin of new forms by transformation from old ones. In the long series of stages which an egg passes through in its transformation into a mature animal one sees not only an actual evolution—the evolution of an individual—but also, more or less obscurely, a repetition of the stages of the long-past evolution of the ancestors of the species.

Man no less than other mammals develops from a fertilized egg, which passes through cleavage, blastula, and gastrula stages. The human embryo has gill slits and aortic arches, which undergo exactly the same transformations that take place in other mammals. Man's heart is at first like that of a fish, consisting of one auricle and one ventricle. His backbone begins as a notochord, is next a segmented cartilaginous rod, then each segment or vertebra consists of five separate bones, and finally each fuses into a single bone. He has in the course of his development three different pairs of kidneys, first a pronephros (or fore-kidney), like that of the lower fishes, then a mesonephros (or mid-kidney), like that of the frogs, and finally a metanephros (or hind-kidney) like that of reptiles, birds, and mammals, which alone survives in the adult. His brain, eye, ear, in fact, all his organs, pass through stages in development that are characteristic of lower vertebrates. Even in those adult features that are distinctively human, such as the peculiar form of the hand and the foot, the number of bones in the ankle and wrist, the number of pairs of ribs, the absence of a tail and the relative hairlessness of the skin—in all these respects the human foetus resembles anthropoid apes more than adult man.

Why are not these and a hundred other structures made directly? Why this roundabout process of making a man? There is no answer but evolution.

Embryology bears indubitable testimony to the truth of evolution; ancestral history is repeated in individual history, but the record is like an ancient palimpsest that has been erased and written over again and again. Traces of the old record are still there; some are obscure, some are almost entirely obliterated, but wherever they are decipherable they tell the old, old story of the common origin of animal and man—their similarities and their kinship.

The Causes of Development and Evolution

Ontogeny recapitulates certain stages and features of phylogeny, but the whole course of evolution through past geologic ages can be followed only by a study of fossils. The record is necessarily incomplete, for we rarely find all the stages of evolution represented in fossils. Nevertheless palaeontology is a certain guide to the general succession of living things during past ages. But evolution is a present process, going on to-day, and to see it at work it is not necessary to explore "the dark backward and abysm of time." To be sure, it goes slowly; species are not made in a day any more than Rome was, but they are being made here and now, and many biologists are studying the steps, conditions, and causes of evolutionary changes in animals and plants. All such study of contemporary evolution is necessarily a study of successive generations of individuals, and all analytical or experimental study of the causes of evolution resolves itself into a study of the factors involved in the genesis of individuals; there is no other possible method of approaching the problem. The study of the factors involved in the genesis of individuals under various conditions of inheritance and environment reveals all that can certainly be known regarding the methods and causes of the evolution of races and species.

The causes of the development of an individual or of the evolution of a species are twofold, internal and external. The internal causes are represented by the organization of the germ cell, the external by surrounding conditions; the internal causes may be called heredity, the external causes environment.

An egg cell, like every other kind of cell, functions in response to stimuli. When a muscle cell is stimulated it contracts, when a gland cell is stimulated it secretes, when an egg cell is stimulated it develops. The stimulus comes, in the first instance at least, from the environment; an egg will start to develop only when stimulated by a spermatozoon or by certain salts or chemicals, or by changes in temperature, and it will continue to develop only so long as environmental stimuli of water, oxygen, food, temperature, etc., remain favorable to its development.

The character of development depends primarily upon the nature (that is, the hereditary organization) of the egg concerned, and secondarily upon the environmental stimuli. The former determines all the possibilities of development and its main course; the latter determines which of these possibilities are realized and modifies more or less the course of development.

Entirely similar causes are at work in the evolution of races or species. With true insight Charles Darwin wrote, many years ago: "Although every variation is either directly or indirectly caused by some change in the surrounding conditions, we must never forget that the nature of the organization which is acted on essentially governs the result." Whether these variations are first wrought in mature organisms and then transferred in some unknown way to the germ cells, as Lamarckians assert, or whether they first appear in the germ cells, as Weismann and his followers maintain, is a secondary, although important, consideration, into which we will not enter here. In conclusion it may be confidently asserted that the causes or factors of the evolution of species and of the development of an individual are fundamentally the same.

Development, Evolution, and Religion

What bearings do these scientific evidences as to the origin of individuals and species have on religious faith? It might satisfy our pride to believe that every human being sprang into existence fully formed and armed, like Minerva, from the brain of Jove, but however pleasing such a belief might be it could not be held by sane and enlightened people. We know well that every human being, even the greatest that ever trod the earth, was once a baby, an embryo, a germ cell, and this knowledge has not destroyed our belief in the dignity of man nor in the existence of God.

It pleases many persons to believe that the first man sprang into existence fully formed and perfectly endowed, coming directly from the hand of God by an act of supernatural creation. But such a belief cannot be held by enlightened persons who have really studied and appreciated the evidences of man's evolution. Such persons know well that every human being bears in his body the marks of his animal origin, and that the human embryo shows that man's ancestors were once water-breathers and later hairy quadrupeds before they became men; yet this knowledge need not destroy belief in the dignity of man nor in the existence of God.

It is a curious fact that many persons who are seriously disturbed by scientific teachings as to the evolution or gradual development of the human race accept with equanimity the universal observations as to the development of the human individual. The animal ancestry of the race should be no more disturbing to philosophical and religious beliefs than the germinal origin of the individual, yet the latter is a fact of universal observation, which cannot be relegated to the domain of theory and which cannot be successfully denied. If we admit the fact of the development of the entire individual from the egg, surely it matters little to our religious beliefs to admit the development or evolution of the race from some animal ancestor; for who will maintain that a germ cell is more complex, more perfect, or more intelligent than man's nearest relative in the animal world?

If the idea of the evolution of a species is atheistic, as some persons assert, so is the idea of the development of an individual, for individual development involves the same principles as race evolution. If one concedes the fact of individual development according to natural laws and without supernatural suspension of those laws, one might as well concede the fact of evolution without supernatural creation, so far at least as its effect on theology is concerned. It is surprising that the so-called "Fundamentalists" have not denied the development of the individual as they deny the development of the species; if they are consistent they will demand that we return to the teachings of the "preformationists" of the eighteenth century—to the idea of endless encasement of one generation within an earlier one and hence to the special and supernatural creation of every child of Adam in the creation of Adam himself. When that comes to pass, there will probably be a demand that the teaching of embryology shall be abolished in all schools and colleges.

How much truer and better is the view that God made the first man as he has made the last and that divine power and wisdom are shown just as fully in the development of the last human child as in the origin of the first! The actual facts of development are no less wonderful than any conceivable acts of creation—indeed they are vastly more wonderful than any that were ever conceived in prescientific times. Just as astronomy and geology and physics and chemistry have given us grander views of the universe than were ever dreamed of before, so biology, and especially the study of development and evolution, have given us grander views of the living world—its unity, its antiquity, its mystery—than were ever before held or suspected.

Among the greatest and most astonishing discoveries in the modern scientific world is the fact that the story told by the gradual development of the embryo gives a summary of the rise and development of its race. A story covering millions of years, if told in a few months or days, must necessarily be very abbreviated, condensed and modified, but the general lines of the two stories agree.

Another astonishing fact shown by the development of the embryo is that it follows the same line of ascent that is shown by the story told by the fossil rocks, thus doubly confirming the fact that the course of nature is from the simple to the complex, as from amoeba to man; that things in nature have come about by gradual change and development, instead of finished and perfect in the beginning.

FRANCES MASON

SADIE

I spent over twenty years in the horse packing business in Wyoming, Idaho and Montana, and have had horses, pigeons, all sorts of wild-life, snakes and dogs most of my life... I could tell a lot of stories about my relationship with animals.

One however stands out. I had a mare by the name of Sadie. She was just plain onery. I or anyone else continually had a hard time catching her - even with a bucket of oats she would always stand off... Once you did catch her and tie her up you were really lucky if she didn't destroy whatever you tied her to. But I'd had her quite a while and I didn't want to send her to the auction for dog food. She hurt her knee one day while bucking a dude off so I retired her to raise some young... She raised some good colts and one filly who turned out just like her mother. Sadie's knee got worse and worse so I had to take her to pasture away from all of her friends. For several months she had been alone and seeing no one or no other horses. When the hunting season ended in the Beartooth Wilderness near Yellowstone National Park, in Montana I decided to go down to Sunlight Basin, in Wyoming and see her.

When I got there she was in a large corral and was in good shape except for her knee and it looked bad. But as usual she didn't want me trying to catch her or get close to her. I stood there and was feeling sorry for her and kind of choked up and got tears in my eyes. She was standing about fifty feet away from me and gave a short whinny and shook her head... And then she walked over to me and nudged me with her nose and made another almost silent whinny. Then she looked me right in the eye. My vision was clouded by the tears and she nudged me again. I don't think I've ever been that close to any part of creation other then Humankind, and I'm sure she knew what it was all about. I'm positive Sadie is a part of Life's Spirit too. Otherwise why would anyone spend millions of years developing her.

BILL HUTH

THE GEOGRAPHICAL DISTRIBUTION OF ANIMALS

Possible Explanations of Geographic Distribution

Different kinds of animals are found in different lands or in different seas. Even the animals of the same continent may show great differences, such as those between the animals of the Canadian forests and those of the coast of the Gulf of Mexico. A hasty examination of the facts might lead to the conclusion that animals were spread over the earth altogether in adaptation to climatic conditions, but this would be a mistake, for climate is only one factor in a very complicated problem, and similarity of climate in widely separated lands is insufficient in itself to bring about similarity of animals. The tropical parts of Australia, Africa, and South America have very similar climates, but their animals are altogether different. Climate, however, may be an effective barrier to the spread of animals and plants, but its action in this respect is entirely negative.

Two alternative views concerning the origin of new forms of life, animal or vegetable, have been presented. One, the older view, which generally prevailed until the publication of Darwin's *Origin of Species,* in 1859, was that each kind of animal and plant had been separately created and was, within certain narrow limits, unchangeable and immutable.

The other view is that offered by the theory of evolution—that all living things now in existence have arisen by natural descent, with modifications, from ancestors that might be traced back, step by step, to unknown simple forms of life.

By its very nature, the hypothesis or doctrine of special creation was helpless to explain the facts shown by the distribution of animals and plants on the earth. On the other hand, if the new theory failed to offer a satisfactory explanation of these facts it would thereby be shown to be inadequate and improbable.

Geographic Distribution in the Light of the Evolutionary Hypothesis

If the evolutionary hypothesis is true, then the present arrangement of living things on the surface of the earth, strange and inexplicable as some of its features may seem to be, must have been the necessary outcome of the whole vast series of changes—geographical, climatic, and biological—through which the earth has passed during unimaginably long periods of geological time. And it may be said, in anticipation, that the present distribution of animals is, in a broad and general sense, explained by what we know of the earth's history as it is revealed by geology and palaeontology. Where explanation is lacking it is invariably due to our ignorance of parts of that history, and we may reasonably expect that the solution of any problem concerning a particular animal or group of animals will be found when all of that history shall have been deciphered. A great many new facts have been discovered since Darwin wrote, and these, as a whole, are strongly confirmatory of the *hypothesis of evolution.*

The mode of procedure followed in solving such a problem may be best illustrated by a single concrete example, which is furnished by the camel family. This very peculiar family consists of two sections. One section includes the true camels of the Old World, whose present native habitat appears to be Asia, for the only wild camels now in existence are found solely in central Asia. The other section includes the much smaller, lighter, and more graceful guanacos, llamas, and like animals of South America. No one who has studied the anatomy of the camels can doubt for a moment that these two sections of the family are closely related, despite the great differences in their external appearance. The geological history of the family, as revealed by fossils, shows that, for a long time during the Tertiary period, North America was its only home, for during that time the camels appear to have inhabited no other continent. Each successive group of rocky strata in our western plains has yielded remains of its own characteristic types of camels, whose development toward the modern type may be followed through many almost imperceptible gradations, all presumably arising by ordinary procreation. Later in the Tertiary period the fossils record the arrival of camels in Asia, on the one hand, and in South America on the other; and finally, at a very late geological date, they completely disappeared from North America. Their passage from North America to Asia was made possible by the existence of land connection where we now find the shallow Bering Sea. This connection was often made and broken in past ages.

According to the doctrine of special creation the relationship between the true camels and the llamas is not real but purely ideal. The fossil camels, which seem to record successive steps of development, all moving in the same direction, record only successive, disconnected acts of sudden creation in a way of which we have had no experience, and the present geographical arrangement of the members of the family bears no relation to its recorded geological history. Such an explanation seems to be altogether incredible, whereas the evolutionary explanation seems to be entirely adequate.

Extinction of Faunas

A great extinction of quadrupeds, which exterminated them over more than three-fifths of the land surface of the globe soon after the appearance of early Man in Europe, is an unexplained mystery. Nobody can yet say why there was such immense mortality among the huge and strange creatures that had roamed over all the continents. Whatever may have been the agent of destruction, it removed from North America a great variety of types, such as elephants and mastodons, many kinds of horses, bisons, tapirs, peccaries, camels and llamas, huge ground-sloths and giant armadillos (immigrants from South America) as well as giant wolves, great lion-like cats, and sabre-tooth tigers. It was this great extermination that put a gap of thousands of miles between the Asiatic and the South American section of the camel family by destroying them in the intervening areas.

Principal Areas of Geographic Distribution

It has long been customary to divide the lands of the earth into zoölogical regions in accordance with the animals that inhabit them. A map on which these regions are indicated by various tints appears to be a very irrational and arbitrary sort of thing. Few of the regions coincide with the continents. Australia and South America fall each into a single region, and Asia, Africa, and North America belong to two, or even three regions each. The northern part of North

America, about to the 49th parallel of latitude, belongs to a vast region (what is called the Holarctic region), which includes Europe, northern Africa, and extratropical Asia. Most of the remainder of North America, including the Mexican plateau, makes up a separate region—the Sonoran. The hot lowlands of Mexico, Central America, and the southern tips of Florida and Lower California are included in the same region as South America (the Neotropical region), as are also the West Indian islands. The regions thus designated are determined primarily by the distribution of mammals, the warm-blooded quadrupeds, for these animals are better known than almost any other group except the birds, and their geological history has been much more fully and minutely deciphered than that of any other of the higher groups. These zoölogical regions are really an inevitable result of the many changes, climatic and geographic, through which the earth has passed while mammals were abundant and diversified.

It is practically certain that no group of mammals arose twice independently in unconnected areas, and it is this fact that enables us to trace, by the aid of fossils, the migration of mammals from continent to continent. If a group of mammals could arise independently and more than once the presence of a given group in North America and in Asia would be no indication that those continents were once connected, but if each group arose but once and spread as far as geographic and climatic conditions permitted, then its presence in two areas now disconnected indicates the former connection, direct or indirect, of those areas. The outlines of the zoölogical regions and their geographic relations afford a key to their history and to the manner in which they received their faunas. The complete zoölogical difference of Australia, for example, from any other continent is in itself sufficient to show that Australia has been geographically isolated for a long time, so that mammals of the higher, more advanced groups were unable to reach it.

The fact that zoölogical regions, as pictured on the map, when marked by differences of colour, seem to make a patchwork or a meaningless pattern on the map is due to the long, varied, and complicated changes in the geography of the regions and in their climate as recorded in their geological structure. Land areas that are now united were for ages separated by arms of the sea, and lands that are now separated were formerly united. Not very long ago, as geological time is reckoned, Great Britain and Ireland were joined to each other and to the continent of Europe, and the North Sea was a wide terrestrial plain, which stood not far above sea level. At about the same time the great islands of the East Indies were parts of the mainland of Asia, from which they became separated at different times. Borneo, Sumatra, Java, Celebes, the Philippines, the Japanese group, and the Kurile Islands were once joined to Asia. In the course of the ages Alaska and Siberia have been often connected and as often separated by uplift and depression of the land, and North and South America were brought together by the Isthmus at a relatively late period. Before that period the two continents had long been separated by a broad sea, which covered the site of the Isthmus and most of Central America.

We must also note the remarkable changes of climate to which the rocks and fossils bear unequivocal testimony. During the greater part of the Tertiary period the climate of the earth was mild and genial, a fact indicated by the fossil plants of the Arctic regions. This climate gradually gave way to one that was colder and that culminated in the glacial period, when so many lands were buried under sheets of moving ice, as Greenland is to-day.

This brief outline will show the great complexity of the conditions to which mammals were compelled to adapt themselves and will help to explain their present arrangement on the surface of the earth, but, it must be insisted, this explanation becomes meaningless if we deny the theory of evolution. If each individual species represents a separate act of creation, then there can be no relationship between species other than that of an ideal plan. On the other hand, if species are really related by community of descent, and new forms arise by the modification of older ones, then many of the facts of distribution are simply and naturally explained as due to changes in geography and climate, of which we have such clear and indisputable evidence.

If we study the mammals of South America of to-day we at once see that they fall naturally into two groups. Those of one group are very peculiar and are limited almost exclusively to the Neotropical region; those of the other are closely related to the mammals of North America. Confined to South America and Central America are a group of strange, bizarre creatures (the Edentata), such as the ant bear, the tree ant eaters, the sloths and armadillos, and, if we include the Pleistocene epoch in our purview, a bewildering variety of two extinct edentate groups, the immense groundsloths and the glyptodonts, or giant armadillos, the remains of which have been found buried in the Pampas of Argentina in astonishingly great number. In South America we find also the platyrrhine monkeys and marmosets and a host of peculiar rodents, cavies, agoutis, tree porcupines, the water hog (the largest living rodent), chinchillas, spiny rats, and many opossums, as well as another type of marsupial, which distantly resembles those of Australia. These constitute a most peculiar and characteristic assemblage of mammals, such as are found nowhere else in the world.

Mingled everywhere with these strange tropical creatures

are mammals of northern type, which compose the second group mentioned above. This group includes the tapirs, the peccaries (or wild swine), the guanacos and llamas, many species of deer, cats great and small, wolves, skunks, weasels, otters, and raccoon-like animals, together with North American types of rodents, rabbits, squirrels, rats, mice, and the like. Though some of the animals of this group are obviously related to those of North America and Asia, nearly all are assigned to different species from their relatives that inhabit those regions, many even to peculiar genera. South American wolves, for example, are in many ways peculiar and must be placed in genera not found in other regions, but that they are related to the northern wolves is indisputable, whether we regard that relationship as ideal or real. If each of these species was created separately, their distribution is in no way explained by the history of the groups to which they belong. If, on the other hand, they arose by natural descent from a common ancestry, this history does explain the distribution, and in a most convincing manner.

Now let us go back to the time, in the middle of the Tertiary period, in the Miocene epoch, when North and South America were not connected, a fact demonstrated by the geological record of Central America and the Isthmus of Panama. We are now in a position to make a full and accurate comparison of the quadrupeds of the two Americas at that time because in Patagonia, on the one hand, and on our Great Plains, on the other, we have immense areas of soft rocks, accumulated at approximately the same time, which have in both the Americas yielded large numbers of well-preserved fossil mammals. The separation of the two continents at that time is reflected in the complete difference of their mammals; in their mammalian life North and South America had literally nothing in common. Miocene Patagonia had no animal that can be regarded as an ancestor of any of those that are put in the second category—animals like those of North America and the Old World. All the mammals of that time and region were either such as gave rise to the peculiar South American forms of the Pleistocene and Recent epochs or such as died out without leaving any descendants. There was a great assemblage of hoofed animals, but they were peculiar, unknown from any other part of the world, and all are extinct and left no descendants in the modern world. There were predaceous creatures, beasts of prey, but no members of the order Carnivora, for these ancient flesheaters of Miocene Patagonia were marsupials, very like the so-called Tasmanian wolf (*Thylacynus*) and related to the opossums. Of the modern cats, wolves, skunks, and other northern carnivorous animals there was no trace; nor of the tapirs, peccaries, llamas, or deer of to-day. There were rodents in great variety, but they were all of the peculiar South American kinds; of the northern rats, mice, squirrels, and rabbits there was not a single representative.

In contemporaneous (middle Tertiary) North America there was an equally rich and varied mammalian fauna, but one totally different from that of the southern continent and very like that of the Old World. In addition to certain characteristically North American groups, not yet known from any other region, there were ancestral types of elephants, rhinoceroses, horses, tapirs, peccaries, deer, antelopes, camels, cats, sabre-tooth tigers, wolves, weasels, raccoons, rats and mice, squirrels, marmots, beavers, hares, and rabbits. The assemblage is essentially that of the Old World, though it shows certain local differences, and it is completely unlike that of South America.

About in the middle or perhaps in the later part of the

Miocene epoch of the Tertiary period Central America and the Isthmus of Panama were raised above the sea, and North and South America were thus connected. This uplift made possible the migration of mammals in both directions, the northern forms going south, the southern forms going north. The earliest northern mammal yet discovered in South America is a raccoon-like carnivore, found in the upper Miocene of Catamarca, the Andean province of Argentine; and the first known Neotropical creature to arrive in North America was a ground sloth, found in the Miocene rocks of Oregon. From this beginning, the proportion of mammals common to both continents steadily increased, reaching a maximum in the Pleistocene, before the beginning of the great extinction already mentioned. In that extinction both continents lost a large proportion of their mammals. In particular, the southern invaders of North America, which had been so abundant in the Pleistocene, nearly all disappeared, leaving only the Canada porcupine as a remnant of that invasion.

In marked contrast to this failure of the Neotropical animals to establish themselves permanently in North America was the success of the northern immigrants in maintaining their foothold in South America. True, many of them, such as the mastodons, horses, short-faced bears, and sabre-tooth tigers, eventually died out, but they died out also in North America, having been among the many victims of the great Pleistocene extermination. However, large numbers are still there and constitute the second group of Neotropical mammals enumerated above. Some of these immigrants, notably the deer and the beasts of prey, have been so long in their southern home that they have undergone considerable modification and must be placed in genera different from those that include their relatives in North America. Among these modified immigrants are the little brocket of Chili and the Pampas deer of Patagonia, whereas the deer of the Guianas is a later arrival and differs but little from the deer of Florida. The wolves, bush dogs, skunks, coati mundis, etc., are obvious variants of northern types. Even the lack of certain animals in North America is reflected in South America, as, for instance, in the almost complete absence of bears, which reached North America from the Old World at a very late period.

Before the geology of the Isthmus of Panama was known, Messrs. Jordan and Evermann made a comparative study of the sea fishes on both sides of the Isthmus. The difference was so great that these authors concluded that the two seas had been separated by the upheaval of land in the middle of the Tertiary period (Miocene epoch), a result which was exactly confirmed by subsequent geological examination.

The geological and palaeontological history of North and South America in Tertiary time is known in greater fullness than that of most other continents, and it explains in a very satisfactory way the existing distribution of mammals in the Western Hemisphere, but only to one who believes the evolutionary theory. Otherwise, that history has no meaning or application, for the existing species are different from those which we find entombed in the rocks, and if they were not descended from the more ancient ones but created separately, then the history has no relation to the present arrangement of the animals. Can any one really believe that successive acts of creation were deliberately arranged so as to produce a false and illusory effect? Absurd as it may seem, such a belief is involved in the acceptance of the doctrine of special creation.

North America and Asia have been repeatedly connected and disconnected at the point where Bering Sea and Strait are now found. Here also many migrations back and forth between the two continents may be traced by the record of the rocks, and these migrations and counter migrations took place at widely separated geological dates. Many American mammals are the modified descendants of migrants from the Old World. Some came to America at a very ancient time; others came recently and are so little modified that naturalists disagree as to whether they should be assigned to European species or not. In short, our understanding of the distribution of existing mammals is dependent upon our knowledge of their history, and where that history is known the distribution is self-explanatory. If the doctrine of special creation is true, this knowledge is illusory, for according to that doctrine the modern species were separately created and therefore have no connection with the ancient species that look so deceptively like the ancestors of the living forms.

North America was separated from Asia at a very late date, as geological time is reckoned, and hence there is relatively little difference between the animals of the American and the Eurasiatic division of the Palaearctic region. The only large mammal that is found in the American and not in the Eurasiatic division is the musk-ox, but until a very recent period that peculiarly Arctic creature dwelt in Siberia and roamed as far westward as Great Britain, which was then a part of the mainland.

The distribution of mammals on the continents finds a satisfactory explanation in the theory of evolution and only in that theory, but the facts of island life furnish even more remarkable and more convincing testimony of the truth of that theory. Islands have been formed at different dates, some rising from the sea within the last few years, others having a history that goes back to the remotest geological antiquity. It is customary to group islands into two classes, suggested by their mode of formation—the continental and the oceanic. Continental islands are detached portions of continents, from which most of them are separated by shallow seas. They have the geological structure of the continents and are composed of the familiar stratified rocks, such as sandstone, slate, and limestone, and some contain granite-like igneous rocks and volcanic material. The faunas of these continental islands are determined by many factors, of which we may disregard several, such as area and climate, and consider only the distance of the island from the mainland and the geological date of its separation.

Great Britain, Ireland, and the islands of the East Indies or Malay Archipelago are continental islands that lie near the parent continents, are surrounded by comparatively shoal water, and although they were detached from the mainland at different times, they were yet, on the whole, of relatively recent origin as islands. Great Britain is, zoölogically speaking, indistinguishable from an equal area of the European continent; the species are so generally identical that the islands must have been separated from the continent during the present geological epoch, the Recent. The great Asiatic islands, Borneo, Sumatra, Java, etc., were detached somewhat earlier and contain more peculiar and more characteristic species, but the difference from those of Asia is not great.

Oceanic islands lie far from any continent and rise abruptly, with steep submarine slopes, from profound depths of the sea. Most oceanic islands are composed of volcanic rocks and coral reefs. Nearly all of them seem to have been submarine volcanoes, which have built up their cones from the sea-floor; and the cones that lie in warm seas are generally capped with coral reefs, which may or may not bury the volcanic pedestal out of sight. The rocks that form the continents and continental islands—the sandstones, limestones, shales and slates, and the coarsely crystalline rocks, such as granite—are not found on the oceanic islands, whose history was radically different from that of the continents and their islands.

When we turn to the animals and plants that inhabit these remote islands we immediately note their marked difference from those of the continental islands. Oceanic islands have no land mammals other than bats, which are carried for great distances by storms, just as land birds are. On some of these islands mice are found, but these may have been introduced by men. Amphibians (that is, frogs, toads, newts, salamanders, and the like) are lacking, because not only are these animals unable to endure sea water, but sea water is fatal also to their eggs. Such islands have no true fresh-water fishes, crustaceans, or mussels. The fresh-water fishes and mussels that they do contain are those which readily enter streams from the ocean and so are much the same everywhere. The animals that are found in these remote islands are land birds, lizards, snails, and insects—such creatures, in short, as may reach them more or less accidentally, as it were, being carried by wind or floated on driftwood for long distances by ocean currents. Their plants also are of the kinds whose seeds are carried by wind and wave.

These oceanic islands may have received their plants and animals in one of two ways. The doctrine of special creation maintains that these plants and animals were directly created where we now find them. If we hold that they were specially created on each island we should expect to find on each island such forms as were particularly adapted to live on it, to which, in short, the environment was suitable and favourable. But this is not at all the case. Many of these oceanic islands are large and could support rather large animals, and the fact that the absence of such animals is due to their inability to reach the islands, except by human aid, is shown by the oft-repeated results of artificial introduction. Rats, mice, rabbits, swine, goats, cattle, and horses, when allowed to run wild on such islands, have thrived and multiplied exceedingly, showing that no unfavourable factor in the environment prevented their presence.

The fact that the oceanic islands really have been supplied with their plants and animals by the more or less accidental agency of the sea and the winds is indicated by a remarkable instance, of which we have the whole history, for it took place within recent years. In the Straits of Sunda, between Java and Sumatra, there is a volcanic island called Krakatau, which, in the summer of 1883 was the scene of a series of the most tremendous volcanic explosions that have occurred within historic time. In consequence of this terrific outburst, most of the island was destroyed and such parts of it as remained above water were so deeply buried by volcanic ash that all animal and vegetable life was completely destroyed; apparently not a living thing was left. Yet in twenty years, or less, the desolate island was restocked by the winds and ocean currents. The late Professor Selenka, of Munich, visited Krakatau some years after the catastrophe, and wrote of it:

> Under the shade of a *Casuarina,* among cocoanut palms and thickets as high as a man's head, I found, to my astonishment, an active animal life of spiders, flies, bugs, beetles, and butterflies; even lizards half a yard in length animated the peaceful picture. All these plants and animals were brought hither by wind and water from Java and Sumatra, replacing the vanished world in the course of a few years.

If the living things on oceanic islands, animal and vegetable, were not created especially for them, they must have been carried to them by winds and currents. If this is true, the nature of the plants and animals of such an island must be determined by the antiquity of the island, its remoteness from land, and its position with reference to strong winds and marine currents. But even if these plants and animals reached the islands in the accidental manner described, might they not have remained unchanged and immutable? In that case there should be no particular relation between the remoteness of the islands and their geological date of formation, on the one hand, and the kind of creatures that inhabit them, on the other. But if species are mutable and subject to modification, then we ought to find that the islands had species that are peculiar to them, yet that would show more or less distinct relationship to those of the mainland from which the islands received the ancestors of their peculiar species. That this is the true solution of the problem is strongly indicated by the plants and animals of the Galapagos Islands, which first led Darwin to form his new views on the origin of species.

The Galapagos Archipelago is a group of five relatively large and ten small islands, all of volcanic origin, which rise steeply from great depths of the ocean. The one nearest to the coast of Ecuador is about 600 miles distant from it, and the islands lie almost on the Equator, in the zone of calms, in which strong winds seldom blow. The arrival of a new form from the mainland must be a very rare event; yet the islands contain many birds, reptiles, and insects, but no mammals. It is fortunate for our inquiry that no aborigines settled in the islands, which, when Darwin first visited them, were almost in a state of nature. Nearly all the animals and plants that inhabit the islands are peculiar to them; the species and many of the genera are found nowhere else in the world. This statement does not, of course, apply to the sea birds, which cross great stretches of the ocean with ease. Though the land birds belong to peculiar species, and most of them to peculiar genera, they are nevertheless of unmistakably South American types and belong to South American families. Large lizards, one a land species and the other marine, as well as the huge land tortoises that have given their name to the islands, are abundant. When the islands were first discovered there were fifteen species of these monstrous tortoises, each island and islet having its own species, but many of these have been extirpated. The species of land birds, like those of the tortoises, are peculiar to a single island each; the genera are mostly common to the group, as the islands are so far apart that communication between them is difficult.

On the theory of evolution these remarkable facts are easily explained. The ancestors of the existing birds and reptiles came, rarely and at long intervals, from the mainland of South America, and after settling in the islands became slowly modified, so that they were placed in genera nearly allied to those of the continent, yet different from them; and, in the isolation of the individual islands the species were free to develop into new forms peculiar to each.

In the Galapagos we witness the results of what may be called a great evolutionary experiment, under conditions unaffected by human interference, and such conditions are rare; but the Cape Verde islands, in the Atlantic, display very similar relations in their animals and plants. The species there are peculiar, but their affinity is as unmistakably African as that of the Galapagos species is South American.

Bermuda and Madeira are almost as far from the North American and African coasts, respectively, as the Galapagos Islands are from Ecuador, but they have hardly any peculiar birds, because they lie in the track of storms, and every year birds are arriving from the mainland in such numbers as to prevent the isolation of groups of them and the development of new types. This must not be understood to mean that new species do not arise on the continents; it merely means that they develop more rapidly under favourable isolation.

Islands like the Hawaiian group or St. Helena, which are very remote from any land and are of considerable geological antiquity, have birds and other land creatures so peculiar that their ancestry is obscure, or even unknown, so that it is difficult to decide what continent their ancestors really came from. This relation between the peculiarity of its species and the remoteness and antiquity of an island is just what we should expect according to the theory of evolution, but these factors should have no effect according to the theory of special creation.

An interesting illustration of this principle is afforded by the rails (Rallidae), a family of birds that is distributed all over the world, on continents and islands, except in the polar regions. The continental species can fly; many of the insular ones cannot. As the islands on which these flightless birds are found were never connected with any mainland, the advocate of the theory of special creation must hold that the birds were separately created on each island or reached it by crossing the sea; but they could not cross the sea by swimming, for no bird is able to swim across such breadths of sea. The birds must therefore have reached the islands by flying and have lost the power of flight after they settled in their insular homes. This loss of flight involves so great a modification of structure that the birds are assigned to new species and even to new genera, different from those to which their flying ancestors belonged. There is no doubt a close relation between the loss of the power of flight and a small, insular habitat. The absence of dangerous enemies, which prey upon the birds and upon their eggs and young, makes flight less essential to their existence. Very often, too, the prevalence of violent winds makes it advantageous for both birds and insects to remain on or near the ground, and not to attempt high or long flights, which involve the risk of being blown out to sea. Many birds do cross the sea for long distances when carried away by storms, but when land birds are so swept from the land they are likely to be destroyed. Most people who have taken a sea voyage have seen land birds, far out at sea, come aboard the ship, exhausted from their long flight, to rest in the rigging. Occasionally these waifs find new homes in remote lands, which have doubtless in this way received their bird inhabitants.

In the Galapagos there is a flightless cormorant, which lives by fishing in the sea. The penguins, those remarkable birds whose wings have been converted into swimming paddles or flippers, live on islands in the seas of the southern hemisphere, where they are safe from the attacks of enemies. The extinct dodo was a large, flightless pigeon, which lived on the island of Mauritius until it was exterminated by sailors and by introduced pigs. Another flightless pigeon, also extinct, was the solitaire of the Isle de Bourbon. New Zealand had many very large flightless birds, the moas, which were destroyed by the Maoris when they settled the islands; and the flightless, almost wingless, little kiwi (*Apteryx*) still lives in New Zealand. It is true that some flightless birds, such as the African ostrich and the South American rhea, live on the continents, but these are large and strong birds and very swift runners, and are therefore able to escape the large beasts of prey and to defend themselves against the smaller ones.

Much the same considerations apply to insects. In Madeira and in Kerguelen Land, a small group of volcanic islands in the Antarctic Ocean, there are large numbers of insects that are unable to fly. Madeira has 393 species of insects that are not found elsewhere, of which 178 species are incapable of flight. These species could not have reached the island in their present flightless state.

When all the facts here presented are taken together—the facts of the animal life of the continents and of the islands—it must appear to any unprejudiced mind that the evolutionary theory offers by far the most probable explanation of them. The alternative theory can offer no solution of problems of geographic distribution. If the theory of evolution were false it would surely be in conflict with the facts of the geographical distribution of plants and animals. When the distribution of a group is inexplicable by the theory of evolution we find that we have not yet deciphered the history of that group. After its history is made known its present distribution is manifestly the inevitable result of a natural sequence of events.

We have already deciphered much of what may be called the geographic history of the earth—the history of the many gradual changes that have taken place in the form and the extent of its lands and seas, as well as in its climate—and it is these changes that have determined in large part the distribution of its animals. All the facts discovered show reasonable natural succession; nowhere can we find evidences of sudden creation; and all are simply explained by the theory of evolution.

WILLIAM BERRYMAN SCOTT

THE HOLY BIBLE - THE NEW TESTAMENT

The General epistle of James 4:4 Ye adulterers and adulteresses, know ye not that the friendship of the world is enmity with God? Whosoever therefore be a friend of the world is the enemy of God.

HOW COULD THESE WORDS BE PRINTED IN A BOOK DEDICATED TO MANKIND?

BILL HUTH

ISLAND BOUND

"It is an interesting fact that on oceanic islands far removed from continents only those forms of life are found which could be borne to them by wind or wave. Only such birds as can be carried long distances by strong gales appear. The fauna of such islands contain no mammals except bats, and in every instance the life, both of plant and animal, is similar to that of the nearest mainland, yet differs from it in having distinct species. If special creation accounts for those forms, why are they not identical with those of the mainland? There is no answer. But evolution affords a simple and inevitable explanation. And if we admit that the original forms of life come from the mainland, and have since changed into new species, then the case of evolution is established."

F. L. DARROW

THE RECORD OF THE ROCKS

WHEN the celebrated Huxley was near the beginning of his career he was very cautious about accepting evolution, and, among other wise warnings, he said that he saw no evidence for it in fossils. At that time Huxley had not specially studied fossils or geology. Later in life he was appointed palaeontologist to the Geological Survey of Great Britain, and he then had studied fossils to such good effect that he was elected President of the Geological Society of London. The more he learned about fossils, the more did he change his early opinion, so that in 1881, when he lectured to the British Association at York, he was impelled to say: "If the theory of Evolution had not already been put forward, palaeontologists would have had to invent it."

The great Swiss-American naturalist Louis Agassiz had a more profound knowledge of certain groups of fossils than any other scientific man of his day. He saw that their distribution in the rocks showed a definite succession and followed certain laws. There was no meaningless scattering—"a tale told by an idiot . . . signifying nothing"—but a history as logical as any that has been written about human affairs. In any group, such as the fishes, which he himself studied, the succession of forms was so orderly and so connected that he even went so far as to speak of their "genealogy" as something that, though not yet worked out, would eventually be discovered. So many of the philosophical ideas of Louis Agassiz are those of the modern evolutionist that it must always seem strange to us that he never accepted the theory in any practical form.

This is a strange contrast—that of Huxley and Agassiz—Huxley, hard, logical, walking on the strait and narrow path, adhering strictly to fact, ended as a champion of evolution; Agassiz, with marvellous intuition, with broad views and wide-ranging imagination, remained its opponent. Huxley, who demanded proof for every hypothesis, was driven to his conclusion by the cumulation of evidence. Agassiz, quite ready to accept an unprovable, transcendental explanation, remained as he was, and the tide of science passed his theories by.

No doubt Huxley in 1881 had a far larger body of evidence to his hand than Agassiz had in 1857; and it may be that Agassiz, had he lived till then, would have found in evolution (though not necessarily in Darwin's explanation) the groundwork of his own metaphysical laws and hypotheses. We today, with half-a-century of additional and indescribably more accurate and detailed knowledge of fossils, can not merely endorse Huxley's statement but can extend and elucidate it.

Notwithstanding, there are people who, having enough knowledge of geology to speak its language, can still deny the evidence. They assert that palaeontologists are arguing in a circle, and their own argument is somewhat as follows: William Smith the land-surveyor, who is known as the Father of English Geology, observed (so they say) a few strata or layers of rock in part of that little island of Britain and found therein a definite succession of fossils. Thereupon geologists jumped to the conclusion that the same succession held good everywhere else. If they discovered that it did not, but that the order was reversed, or that there were great gaps, then they explained such exceptions by saying that the rocks had been overturned or that large portions of them had been removed, and so on. On this assumed succession of the fossils the palaeontologists based a number of lines of descent of the extinct animals and plants and claimed that the changes were due to evolution. Links were admittedly missing, but sometimes a fossil was found that fitted into one of these supposed breaks; then the palaeontologists put it in and called it fresh evidence for evolution; and if the geological age did not quite suit their theory, they said the geologists were wrong, and that there must have been some disturbance of the rocks. In short, the theory was based on the succession of fossils in the rocks, and the succession of the rocks was deduced from the theory. A vicious circle if ever there was one!

So far the critics, but it is somewhat difficult to make out what explanation they would themselves give of the facts presented by the fossils. They do not deny that fossils are the remains of extinct animals. They seem to suppose that the ringed trilobites, the coiled ammonites, the armoured fishes of the Old Red Sandstone, the scaly bony fishes of the Chalk, the monstrous dinosaurs, the huge horned mammals, the great marine ichthyosaurs and plesiosaurs, and hundreds of other forms unknown to us today, all lived at the same time, though in different regions or different situations, and that the rocks are a sort of hotch-potch in which their remains occur anyhow.

This explanation, or any other conceivable interpretation of our critics' views, only raises more difficulties and is hopelessly inconsistent with the most readily ascertained facts revealed in any quarry or coast section. But until it is put in clearer form it is not worth discussion. It will be more to the point to see what really is the nature of the argument which these critics travesty.

When I was at school at Winchester, which is one of the oldest cities of England, new drains were laid down and a deep trench was dug along a street in ground that had never been disturbed before. (Fig. 1.) Passing this trench every

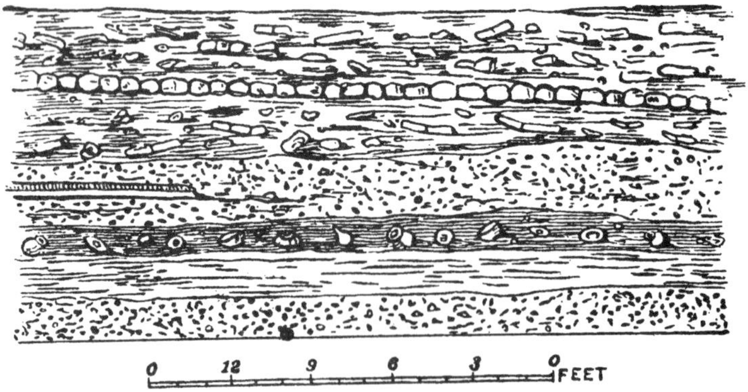

FIG. 1.—Section exposed in a city street.

day, I acquired many articles that the workmen threw out. Just below the latest road metal they found coins of the Georges and fragments of china; at a lower level they turned up a bit of green glazed mediaeval earthenware; below this, among other Roman remains, was a piece of the red ware known as Samian; still lower was a fragment of the rough black earthen pottery made in Britain before the Romans came. These objects, which I still possess, with various coins and other things, were all in such an order that the historically oldest lay at the greatest depth. From a layer below these I obtained a piece of hard bone, shaped before the age of iron into a stout pin; and below this again was found a fine stone implement.

No one can doubt that in this old street of Winchester the ground had gradually risen layer by layer, and that each layer contained objects dropped by successive generations of men who lived while it was forming. From these facts alone one could learn much of the history of Winchester and could reconstruct that succession of cultures which, as we know from other facts (not to mention written documents), is applicable to at least the whole of England. These coins and pots and tools are fossils, and their orderly suc-

FIG. 2.—Diagrammatic section across the London basin.

cession is everywhere the same, unless, indeed, the ground has been disturbed by subsequent building operations.

If we dig deeper we shall come, it may be, to layers of gravel and brick-earth, as we do in London (Fig. 2); then to stiff clay, and below that to other harder rocks. Except perhaps in the upper gravels, we no longer come across the remains of man, but we find the bones and the shells of other animals, and these, we note, occur in just such regular succession as did the coins and pots. Are we not bound to make a similar inference and to say that the fossils indicate successive layers of rock and a succession of animal inhabitants? We do indeed find that the more closely we study any thickness of rock the more does each successive layer prove to contain its characteristic fossils. These layers can be seen and measured and traced across country, and so far as they have been traced their succession is the same. When we come to a gap (say the English Channel) and when we find that the succession of fossils on the other side is the same, must we not suppose that the succession of the rocks (say in France) is the same as in England?

In this way, step by step, it has been possible to pass from one country to another, until the stratified rocks of the whole world have been correlated or linked up just as the succession of Emperors in China can be correlated with the Kings of England. There are difficulties, no doubt, owing to large gaps like the Pacific Ocean, or to the readily intelligible fact that some animals live in the sea and others on land or in fresh water; also to the fact that there are differences between the inhabitants of arctic and tropical climates. But, so long as attention is limited to fossils of similar nature, these difficulties rarely obtrude. One finds, for example, in the State of New York and in Quebec a clear succession of fossils precisely comparable to the succession in Great Britain.

It is perfectly true that there are places where a rock (A) overlying another (B) contains fossils that elsewhere are found to lie below B. But we shall generally find that such places show signs of disturbance of the rocks. If we admit (as everyone does) that the fossils found in the Alps, the Rockies, and the Himalayas are the remains of sea-animals, then we recognize that these mountain chains must have been raised from the sea-floor by forces so enormous that they could not help crushing, crumpling, and overturning the rocks. There are plenty of obvious evidences that this has happened, and it may well have happened in places where the signs are not so obvious. Tremendous though these movements seem by our pygmy human standards of comparison, it is well to remember that the greatest ups and downs on the surface of the earth are relatively less than the irregularities on the skin of an orange. Now picture to yourself the rocks of the Alps before they were folded and raised, and you will see that they must have occupied a greater area, as a folded tablecloth does when you smooth it out. Thus you can appreciate that some of these masses of rock have been shoved many miles and thrust over other masses. Naturally, in such a process the original succession has been disturbed.

Now the evidence for these movements does not entirely depend on the fossil remains of living creatures. Such folding and thrusting can be followed in the rocks of Finland, which contain no fossils at all. But the presence of fossils in the rocks, if we admit the evidence from undisturbed areas, is of much help in working out the succession of the disturbed rocks. The evidence is of just the same nature as that on which we rely in tracing the history of some ancient cathedral that has often been partly pulled down and rebuilt.

The evolutionist does not base his conclusions on evidence from these disturbed areas, where indeed the fossils are too often shattered and obscure. He is content to take the far larger areas of the earth, where the succession is clear and the rocks are only a little tilted. Wherever in such a succession we are able to trace the history of a single group of organisms, we find it perfectly continuous and regular. Gaps there may be, but the more we explore and study the fewer are the gaps. This continuous history always shows a gradual change from the oldest to the newest forms, and at no point is it possible to say that there was an entirely new creation.

It is not easy to find a great thickness of rock that was laid down continuously through many thousands of years.

Almost everywhere there are breaks, due either to some cessation of the supply of rock-material or to the washing away by sudden local currents of rock already deposited. Or again, though rock may have been laid down without pause, the conditions may have changed, so that, for example, sandstone is succeeded by fine-grained clay or shale, and this again by limestone. With the change of conditions there was a change in the character of the animal and plant communities living on the sea bottom or in the waters above. Some creatures live at one depth, some at another; one kind prefers a sandy bottom, another prefers a limy ooze, and so on. Consequently, where a great thickness of rock consists of layers differing in composition, it is not possible to trace a single race of animals up through the whole sequence in any limited region. The race migrated and its descendants must be sought elsewhere.

One of the best examples of a rock that was laid down continuously through a long period of time and yet retained its general character is the Chalk of southern England and northern France. This soft limestone was laid down in a relatively shallow and apparently calm sea, and it contains many fossils of marine animals. Among these fossils sea-urchins are very common, and a careful study has been made of one particular genus of heart-shaped sea-urchin named *Micraster.* A. W. Rowe, a physician of Margate, devoted his holidays to collecting these heart-urchins from the Chalk of England, foot by foot. He was able to show that what appeared to be a distinct species found at the bottom of the Chalk gradually changed into a different species found at the top. This change is almost imperceptible, but it can be traced in every part of the fossil shell, and it takes place in the same way in all parts of the country. Here is an example of evolution caught in the act. If we were to take a set of photographs of these fossils from the base of the series to the top, and copy them on a cinematograph film we could see evolution taking place before our eyes.

Let us remember that this change of one species into another took thousands of years, probably hundreds of thousands; then we shall not expect to find evidence that similar changes have taken place during the brief span of historical time.

One should note also, both in this case and in others that are equally well established, how very regular is the course of the change. A drunken man staggers along, veering from one side of the road to the other, stumbling and stopping at random. His aim is not visible; his course cannot be foreseen. How different is the flight of an arrow towards its mark; rising from the archer's bow, and then sinking in one gentle unbroken curve till it pierces the bull's-eye! This mathematical regularity is due to the momentum imparted by the bow and to the pull of gravity; any deflections due to the wind can be allowed for and calculated. Of like nature, and no less due to natural causes, is the regular change of an evolving series of animals. But whatever may be its cause, the regularity is such that the palaeontologist can predict the existence of forms still unknown, but required, on his theory, to fill a gap or to extend the series backward. Such forms have often been found in accordance with his prediction. This power of correct prediction is generally held to be the strongest proof of any scientific theory.

For these reasons palaeontologists are bound in honesty to accept evolution; but equally in honesty they must confess that they do not yet know all its laws or all its causes. That they disagree upon what they do not know does not prove their testimony false if they agree upon what they do know.

VOICE OF CREATION

ONE FAR-OFF DEVINE EVENT

"Everywhere is ceaseless change. Nowhere is there perfect adjustment and harmony, always a constant becoming, a never-ceasing growth. The world is never finished, never a flawless work of art, but always striving toward perfection, taking its course, it may be, toward that one far-off divine event toward which the whole creation moves."

F. L. DARROW

Fossil remains show that none of the plants or animals of the distant past were like those of to-day. They prove that species have altered since the beginning; and the farther back in time we find the remains the more imperfect, the series showing a gradual improvement in forms.

FRANCES MASON

GOD OUR PARTNER

If a person has a desire to understand the existence of Eternal Life, he or she would have faith in God's plan through the Spirit of Creation. For millions and millions of years God has guided creation down an easier and more pleasant road of existence; and it gets easier day by day.

A person must feel God in his heart, have understanding with Him and His ways and patiently listen and wait for each of His accomplishments to evolve. He has proven Himself with you and the rest of creation over and over; yet His work will never be complete... His daily dealings are through the Holy Spirit. Everything that is He must guide, and all that have passed through and into this Spirit will assist Him in His plans being accomplished.

There is nothing which you see and that which you do not see that is not a part of our partnership with God in the Holy Spirit. There was no beginning and there is no end.

When you open your eyes you are seeing all creation as God: When you close your eyes you are in the Spirit. All Creation is in the Spirit... Death can be described as a means to be born into the Holy Spirit: Birth means being supplied with a soul to once again, "hit the streets".

BILL HUTH

THE NATURE OF SPECIES

THE final test of the theory of organic evolution is whether the different kinds of animals and plants are fixed and unchangeable or whether one kind may through its posterity give rise to or pass into another kind, even though the passage may be so slow that the changes produced in the lifetime of one observer are slight. No highly specialized organism can be expected to develop into an altogether different organism; there is no chance, for example, that a humble-bee will give birth to an elephant, or that a club moss will develop into a fruit tree. Nor can any decisive verdict as to the natural evolution of new forms of animals or plants be given from a consideration of the various breeds of sheep or dogs or garden plants. New breeds can unquestionably be developed at the will of the breeder; but the fact that domesticated animals can be varied by breeding does not necessarily show that under natural conditions the progeny of one kind of elephant can become another kind, or that a certain sort of moss, if placed in a new environment, will become another kind of plant.

The work done by breeders shows the plasticity of living forms, and like plasticity is seen in animals and plants living under natural conditions, but the question is whether there are in the animal and vegetable kingdoms any well-established units between which distinctions—even comparatively slight distinctions—can be marked by boundaries that are impassable. Such units were once supposed to exist, and they were called species, a word meaning a kind. The name species was given under the impression that, though each kind so designated might vary to some extent, the variations were restricted within limits that could not be transgressed. The differences between members of the same species were regarded as individual variations. Slight differences between plants and animals of the same species were recognized by careful observers, and the abrupt changes seen in sports and monstrosities attracted the attention of the curious. Many early authors placed no limit on the extent to which such variations might occur. Bacon observed a fern growing out of a willow and, instead of explaining it as a natural graft due to a windblown spore caught in a crack, regarded it as an offshoot due to some injury or some special influence. He also suggested that the stump of a felled beech might put forth birch, it being "a tree of a smaller kind which needeth less nourishment." Thus, according to Bacon, a beech might be developed into a birch by an unfavourable environment.

The belief in the fixity of species arose in the generation after Bacon. Herbert Spencer[1] attributed it to a literal acceptance of the Mosaic account of Creation, and it has often been credited to Milton, who relates, in Paradise Lost,[2] how, on the sixth day, in accordance with the Divine command,

> The earth obeyed, and, straight,
> Opening her fertile womb, teemed at a birth,
> Innumerous living creatures, perfect forms,
> Limbed and full grown.

But, as Professor Poulton has remarked, the belief in the sudden appearance of animals in their present forms was

[1] In 1852, reprinted in Essays, vol. I, p. 583, 1868.
[2] Book VII, lines 387-500.

due less to Milton than to the faith of his fellow Puritans in the verbal inspiration of the Bible.

The formal definition or establishment of species was practically begun by Milton's contemporary, John Ray, an Essex naturalist (1627-1705). Ray founded many of the species of the British flora and prepared the way for Linnaeus (1707-1778), whose system was based upon implicit faith in the immutability or fixity of species. Linnaeus declared, in a famous dictum, that "the number of species is as many as different forms were created at the beginning." In the following century Cuvier was equally positive. He believed that species are as distinct as the different makes of boots sent out from a factory. Darwin, on the contrary, called his epoch-making treatise "The Origin of Species," because he maintained that species pass into one another; and the doctrine that one species may be derived from an earlier allied species is the doctrine of organic evolution. He regarded the term *species* as one arbitrarily given, for convenience of designation, to a set of individuals closely resembling one another—as a term not essentially different from the term *variety,* which is given to less distinct and more fluctuating forms.

Darwin's theory was opposed to preconceived opinion. The species of the more highly organized animals and plants appear to be sharply separated. The differences between the African and Asiatic rhinoceros, between the African and the Indian elephant, between the one-humped and the two-humped camel, appear to be constant and absolute. Nevertheless, when such animals are examined carefully it is found that individuals in different herds of each kind show slight but significant differences. The giraffes, which were at first classified as one species, have been broken up into eleven subspecies; the African elephant has been found to include more than one species and several subspecies. Systematic botanists and zoölogists have been divided into two schools—the "Splitters" and the "Lumpers." The "Splitters" establish species on differences which the "Lumpers" treat as mere individual and inconstant variations. Darwin represented Asa Gray, the famous American botanist, as a Splitter, and Sir Joseph Hooker, of Kew, as a Lumper. Herbert Spencer, in 1852, estimated that the number of species must amount to at least ten millions. The Splitters would multiply that number many times.

There has been no agreement as to what characteristics should be regarded as of specific rank—that is, as sufficient to justify a naturalist in founding a species—and as to what are of a lower systematic value. For example, there has been a long-continued controversy whether man is one species or whether the European, the Negro, and the Mongolian are distinct species. This difficulty has been partly overcome in practice by the introduction of minor units of classification, which have been called subspecies, and the subspecies have been divided into varieties, and these into subvarieties, and these in turn into races and subraces. The divisions are thus numerous, and the grounds for them are indefinite. Different groups of plants and animals have different grades of specific subdivision, according to the abundance of their members, or their variability, or the attention they have attracted. Thus the Flora of France uses in some genera six subdivisions lower than the genus. British botanists adopt more subdivisions of species in roses and brambles than in less variable plants. Some species of British land snails, such as the common *Helix nemoralis,* have undergone indefinite subdivision.

The extent to which experts differ as to whether certain variations are distinctive of species, varieties, or races shows that there are no such fixed limits to species as the pre-

Darwinian naturalists believed. A species is an expression of opinion, not of fact.

Persistent efforts were made to delimit species by the sterility of hybrids. The fact that the mule is sterile was set up as proof that the horse and the donkey are different species. But the rule that hybrids are sterile is subject to many exceptions; thus the rabbit and the hare have a fertile hybrid known as *Lepus darwinii,* and so have the common and the Chinese goose, which are classified as unquestionably distinct species. Conversely, crosses between many domesticated varieties of plants are sterile.

The fact that a species is an arbitrary and not a well-defined natural unit is further shown by variations of organisms from the standard types. Herbert Spencer, in 1852,[1] years before the publication of Darwin's *Origin of Species,* claimed that it had already been proved that "any existing species—animal or vegetable—when placed under conditions different from its previous ones, *immediately begins to undergo certain changes of structure fitting it for the new conditions.* They [the supporters of the theory of natural development] can show that in cultivated plants, in domesticated animals, and in the several races of men, such alterations have taken place. They can show that the degrees of difference so produced are often, as in dogs, greater than those on which distinctions of species are in other cases founded. They can show that it is a matter of dispute whether some of these modified forms are varieties or separate species."

Some variations are obviously due to the influence of environment and of mode of living—of daily work. Variations that are successive and cumulative in time have been called mutations (Waagen, 1868), though that term has been

[1] Reprinted in Essays, Vol. 1, pp. 379-80, 1868.

used later (de Vries, 1901) in a nearly opposite sense—that is, to denote sudden variations, or sports.

Herbert Spencer remarked: "Those who reject the theory of evolution as not adequately supported by the facts accept instead a theory which is supported by no facts at all." The development of specific differences—differences marking species—by gradual change from generation to generation has been well established by collecting fossils from successive

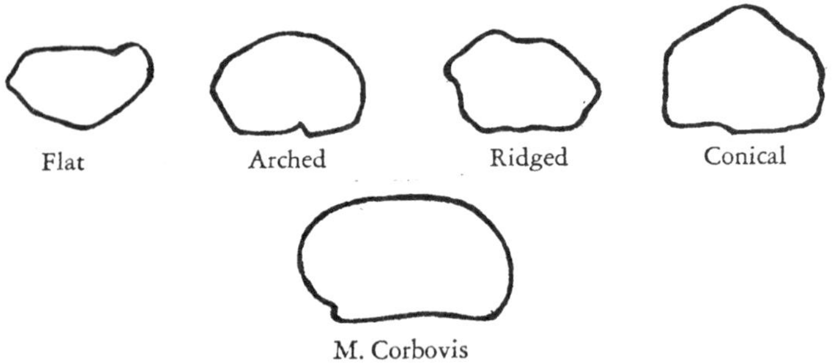

FIG. 2.—Sea-urchins formed in the English Chalk.

layers in a series of deposits and comparing each fossil with its predecessors and successors.

The sea-urchins in the English Chalk that belong to the extinct genus *Micraster* (Fig. 1) provide a convenient illustration of the evidence thus obtained as to the actuality and nature of evolution. This sea-urchin is a common and well-preserved fossil, and it shows variations that might be regarded as distinctive of two or more species. The late Dr. A. W. Rowe, of Margate,[1] collected 2,000 specimens of *Micraster* and carefully recorded for each specimen its level in the Chalk. They show gradual variations (Fig. 2) as they are followed upward through the Chalk. They are of four chief shapes, all of which have come from the earlier

[1] Quart. Jour. Geol. Soc., Vol. 55, 1899, pp. 494-547, pls. 35-9.

form *M. corbovis.* One kind retained the flat top and elongated form of the typical *corbovis;* another became arched above; a third developed a prominent posterior ridge; and a fourth became conical or pyramidal. These four series may all be included in the same species, being, according to Poulton's term, epigonic—that is, descended from one pair of ancestors; or each series would answer to Darwin's conception of a species, for each showed an appreciable difference and persisted for a considerable period. Rowe's collection, however, showed that though the different shapes were developed by slow change, they were not developed by continuous divergence from the original type. The contemporaries of the early *corbovis* included dome-shaped and flat varieties. The forms of *Micraster* found at each horizon or each level in the Chalk show variations that were probably dependent on the nature of the sea floor and on the movement of the sea water. The evolution of the four chief varieties of *Micraster* in the Upper Chalk was not due to steadily progressive variation from the ancestral type but proceeded by innumerable minor irregular variations, the effect of the environment acting upon successive multitudes of *Micraster.* The evolution was doubtless due to syngamy or interbreeding under natural conditions, with the encouragement by the environment of the useful modifications. It did not proceed along lines of continuous and steadily diverging variation, but along several parallel lines, from each of which there were variations that would overlap those from the next line. Dr. Rowe, in his diagram showing the evolution of the Chalk *Micraster,* used the plan of branches divergent like the twigs of a tree. The main types were established by parallel lines of descent, the variations from each line being radial in all directions and overlapping those from adjacent lines. This conception agrees with that stated by Poulton when, in closing a discussion at the Entomological Society on "What is a Species?" he remarked [1] that he had never conceived of the origin of a species from one ancestral pair, but always from the change of masses rather than of individuals. He added that it was "the splitting of the single community into separate subcommunities which was the foundation of the process."

These subcommunities were the kinds of groups for which, in 1896, I suggested the term *circulus*. In a catalogue of the Jurassic Bryozoa, or "moss-animals," in the British Museum I pointed out that the term species is inappropriate to these groups, because they are not separated by definite boundaries and were not developed by continued divergence into isolated assemblages. The term *circulus* was suggested from the analogy between these groups and the knots of people who collected around the speakers in the Roman Forum. Each knot would be crowded near the centre and looser on the margin, whence people would frequently pass to an adjacent circulus.

The term circulus was also used in 1900 in a work [2] describing a large collection of fossil corals from the Jurassic deposits of Cutch, in western India. Most of the corals came from one reef, which was especially rich in the simple coral *Montlivaltia,* of which there were more than 2,000 specimens. Each circulus of these Indian *Montlivaltia* shows variations as great as those representing species among the corresponding European corals. One of the flat corals, named *M. frustriformis* because it is shaped like the frustrum of a cone, has fifteen European analogues; a taller horn-shaped form, *M. cornutiformis,* corresponds to twenty European species, and *M. kachensis* to eleven. In the European

[1] Trans. Ent. Soc. London, 1911, p. XVI.
[2] Palaeontologia Indica, Ser. IX, vol. 2.

Jurassic rocks many of the coral reefs are isolated, occurring at places where warm currents from the south raised the temperature of the water. Most of these coral reefs have disappeared, and thus, though the development of the European *Montlivaltia* may have been continuous, it appears discontinuous owing to the imperfection of the geological record. In Europe it is therefore convenient to treat the separate groups as species; but in India, where a large number of corals were collected from one area of slightly undulating sea floor, the variation is continuous. The groups of corals are knots of individuals, or circuli.

The opportunities for tracing progressive evolutionary series of fossils in the field are not numerous; such series can be found only where thick deposits have been laid down continuously under the same geographical conditions, so that for a long period forms were deposited one above another and the intermediate forms were preserved in their right order. The Chalk is one of the formations to which this method can be applied. It is a soft, earthy limestone, in places a thousand feet thick, and was laid down as an almost continuous deposit of limy mud, so that the fossils are perfectly preserved and easily extracted; and owing to the many uses of chalk large exposures are available in inland quarries as well as in continuous sections in sea cliffs. Other groups of Chalk fossils show the same continuous evolution as *Micraster*. The process has been demonstrated in other formations, as by Hyatt and others for the ammonites, by Carruthers for a Carboniferous coral, by Schuchert and H. Walker for the brachiopods. (See Fig. 3.)

Opportunities for the study of contemporary variations are more common, both with fossils and with living animals or plants. Where organisms live in large numbers under similar conditions the attempt to divide species becomes practically impossible. Shell-fish which are so scarce that a museum collection contains only a few representatives are easily divided into species or varieties; but shell-fish such as oysters, which live together in multitudes, are indefinite and uncertain as to species. The same difficulty has been observed with the sea-butterflies, or pteropods, which live in swarms on the surface of the sea and form a large part of the food of whales. In groups where specific variation was slow, or the members were few, or the fossil remains are rare, the differences are so well marked that the delimitation of species presents no difficulty. Organisms, however, that live together in vast numbers and under similar conditions show continuous variation, and though the individuals may be massed around certain centres, the groups grade into one another.

The arrangement of such groups into circuli instead of into species is a fulfillment of Huxley's prediction in 1880 that "The suggestion that it may be as well to give up the attempt to define species and to content oneself with recording the varieties . . . which accompany a definable type . . . in the geographical district in which the latter is indigenous may be regarded as revolutionary; but I am inclined to think that sooner or later we shall have to adopt it."

The artificial nature of species has been generally recognized by working naturalists; but the term species is still retained. Sir Ray Lankester, with his logical consistency, recommends that it should be abandoned; but it has been maintained from tradition and convenience. The abstracts of the papers contributed by Prof. H. L. Hawkins and Dr. A. E. Trueman to a recent British Association discussion on the "Conception of Species" show that their idea of a species is that of the circulus; and so also is the "species-group" of Dr. Bolton among fossil beetles. It is fully time that the term species should be less frequently used, as it is apt to mislead. It was based on the belief that species are fixed and immutable, and its use encourages that belief. That "species are species" is a statement often made in support of the idea that naturalists in practice treat species as fixed.

The circulus, on the other hand, is a natural grouping, which adopts evolution as a fact and as achieved by slow variation in all the members of a group in various directions. Each circulus has what Bateson called a centre of organic stability, and most of its members tend to be near the centre so long as the condition remains the same. Wherever during the growth of the Cutch *Montlivaltia* the sediment deposited on the sea floor accumulated steadily, most of the corals would have the same ratio of diameter to height; but if it accumulated at one place more quickly than at others, the centre of organic stability at that place would be with corals that were higher in proportion to their width.

The circulus provides a nomenclature which is consistent with the view that evolution results from changes affecting the mass of individuals belonging to a group and which relieves the naturalist of worrying over, say, the number of species among British brambles. Discussions of such problems, except in so far as they stimulate close observation, are comparable in utility with the mediaeval arguments as to how many angels could stand on the point of a needle.

JOHN WALTER GREGORY

VOICE OF CREATION

SIXTY-TWO TO TWO

"All students were so impressed with the belief in the reality and permanence of species that endless labour was bestowed on the attempt to distinguish them—a task whose hopelessness may be inferred from the fact that even in the well-known British flora one authority describes sixty-two species of brambles and roses and another of equal eminence only two species of the same group."

A. R. WALLACE

THOROBREDS

"A species is supposed to be a group of individuals that closely resemble one another owing to their descent from common ancestors—a group that has become more or less sharply separated from all other coexisting species by the disappearance of intermediate forms." "The more we study the animal and vegetable kingdoms . . . the more clearly is the fact impressed upon us that if we could have before us all past and present individuals we should find it impossible, except in an arbitrary manner, to arrange them in species at all, for each kind would be found to be connected with others by series of small gradations."

ARTHUR DENDY

LEAVE IT UP TO BEAVER

"The question what constitutes a species must be left to the judgment or fancy of the individual."

HENRY SPENCER JENNINGS

O SPECIES INCREASE

Modern students of nature do not find, as Linnaeus stated about two hundred and fifty years ago: "There are as many different species of animals and plants on earth as there were different forms created in the beginning."

UNKNOWN

EXTENDUALISM

Before I undertake the explanation of the *word* Extendualism let me first make clear to the reader that they should not mistake Extendualism with the philosophy of Existentialism.

Existentialism being the observance of oneself by taking an outsider's view of what is taking place within. In practice the study, in itself, to prove the existence of the single personality and its continuance through reincarnation.

Extendualism, a *word* hereby derived to fit the situation which has always existed, is taking as fact the existence of the single personalities which have been blended through desire of what has been considered the best for promoting creation as One towards the goal of a Kingdom of Heaven at hand and with the objective in mind to make the individual realize the fact of an existing Inner Spirit which acts as the driving wheel of progress in all life's production. Also using this *word* to express the hypothesis that the individual is moving about in continual reflection of the actions and vibrating sounds of life which surround him; these perpetually being mentally and physically digested in order to calculate that part which will be more beneficial for our X, individual who lives as a part of the makeup of God eternally with God's Spirit. X, feeling that what will be most meritorious for him will in turn be of most value to the surrounding creation in which he exists, and thereby the aspirations of our individuals are to reflect what is felt to be of highest value for the progressive harmony of society as a whole: all X's in their own feeling, and in reality they are, capable of summarizing which choice will be best for his surrounding life and its rhythmic cycles of orderliness-be them Adam or Atom-and then reflecting these in his personal makeup.

Although not clearly understood to himself this X is matter of fact choosing for himself to reflect to the Inner Spirit with

All. This statement being to express clearly that there is but One Spirit-*That* part of *All* life, which all life questions the existence of and which is today interpreted by the *word*, God.

In maximum the *word* Extendualism meaning Oneness with creation: a creation which unceasingly motivates through an energetic process towards an easier and more understanding way in which It, as a whole, can Be.

Although, as before stated, the practice of Extendualism has always prevailed to carry the Spirit forward towards the understanding of Oneness it has only been the lack of a means of expressing it verbally which has suppressed any expounding of the theory to students who seek to reflect clearer views as to the question of Being and its progress.

Pantheism, the doctrine which teaches that the universe, taken or conceived of as a whole is God, could be closely related to Extendualism. However due to the lack of a definite explanation to *The* question, this philosophy, too, has been discarded as another pious fraud.

As you, I too, want to feel myself necessary, and to in some way have a part in future life. To do this I have run the gauntlet of religious teachings-both as a student of philosophy and spiritualism, and to these I have added the now accepted sciences which pertains to Life Energy and physical and mental evolution. It is through the alloy of these plus my observance of Life Itself that I offer, as Mr. X, my hypothesis-Extendualism . . . .

It will be up to you, part of Me, to then digest what seems correct and dissolve it into what you may feel to be a clearer explanation of Life's question and then reflect this into the world of Being to be reworked by other parts of *All* which in turn will further the progress of understanding.

Most prominent in my personal experiences with life Extendualism in mass operation came to me as I was sitting in a small newsreel theater in Hollywood, California. Here, I was witnessing the filming of an uprising in the Far East. To me and my fellow-viewers, it was impossible to discriminate between the *good* or *bad* side of the fighting. The camera was focused on two young men-one firing a rifle at the other who had no weapon. The fleeing youth was finally hit and he dropped to the ground. As the assailant moved towards the fallen victim the cameraman moved in closer. The armed man then raised his rifle and fired at the man who was laying on his side in the sand and who's very young face was filled with fright and begging plainly for mercy. The bullet tore into the skull above the nose bridge and blood instantly gushed from the writhing mouth and from the cavity which was caused by the lead slug. The body, a part of God-You and Me, writhed in frenzied actions and then lay silent in the blood stained sand. As the toll of the action took place the audience in the theater, en masse, gasped and left me as most others, with the knowledge that a part of Life's Spiritual Energy had been wasted. Too, I, as the others, seemed to feel that instant of piercing pain.

The gasp: Extendualism in action.

Under these morbid circumstances are not the only times when a person extends himself to be another. In everyday conversation a person seldom makes remarks to others without first judging the effect it will carry to that person with whom he speaks-in this way, they, the speakers, are extending themselves to be the receivers of the thought before it is spoken-a sort of censoring for the betterment of that other part of themselves.

Also the person who smiles at another's success is in all truth extending himself to be that person: he can justly feel the enjoyment which will be his fellow-parts with success.

How many times have you, personally, seen another person being beaten by an assailant who refuses, in his action of fanaticism, to stop striking his unconscious rival, and who is finally forced by the spectators who intercede and physically restain him from his intolerable act.

The dog, who is ignorantly considered by many as less than man, is also shown this same respect. More for that matter-any part of life will not be allowed to be wasted by another without some sort of intervention.

The Inner Spirit, God in Extendualistic action . . .

Many have there been who were great in the practice of fully extending themselves to be a part of All. Of these who display and teach Extendualism is Jesus of Nazareth. His teaching is filled with those thoughts which are digestible to fit in with the Extendualist's philosophy. Many quotable statements could be used, however I have the privilege, as Mr. X, to use what I feel will carry the greatest force in conveying my message to you. *When you are kind to the meek, you are kind to Me.* In what greater way could a man who had been educated to be different have expressed Himself to be a part of All than to say the *Meek?* For to Him *the meek shall inherit thc earth.* How true . . .

Moses, another devoted part of The Life Spirit, made one of the greatest statements which has come from the lips of mankind. Thousands of years before Jesus, this man spoke to us in order to enlighten The Spirit to the fact that, *whoso sheddeth man's blood, by man shall his blood be shed.* I have used this statement many times in trying to express the relationship of one to another, and perhaps have overdone my anxiety in its use. However I must stand solidly by its meaning. The blood of one is the blood of All-be it red, green, black, white, or colorless: It comes from One as Energy and will be with One for Eternity: it is this One that the Extendualist takes pride in Being.

Try forgetting or removing yourself from anticipating the

effect you have on others. As you move, as you eat, as you sleep, as you dress, as you speak, as you listen, or as you are is automatically extended through uncontrollable reflection to your immediate surroundings who then in an unconscious way, as Mr. X, digest and then reflect the part which they think to be most meritorious of you onward by reflecting through themselves and this action continues to gyrate its course which will continually make it better for you-you, who since the beginning of this article are no longer the same person-neither in appearance nor mental makeup and who, as much as you feel not, are part of-and have Life Eternal through the Spirit Life.

Extendualist *that your are* will bring about the action of *not* narrowing your world and simplifying it to fit your individualism and feeling that because of your devotion to a symbol-like god and sacrifice to the god will promote you to a reincarnation in a heaven which lies beyond this earth, but in fact it makes it necessary to surrender what you feel to be yours, you individualism, to the Spirit of Life. Thus making what you feel to be yours, you, extend into its rightful place: an active part, which in actuality it is, in the makeup of what we, God, call God.

The Kingdom of Heaven is at Hand -- it is only for us to understand. Its extension will then come automatically to embrace All through what you are now plainly and as a truth exhibiting -- *Extendualism* . . . . .

BILL HUTH

VOICE OF CREATION

I'M WITH YOU ALWAYS

Do not stand at my grave and weep.
I am not there, I do not sleep.
I am a thousand winds that blow,
I am the diamonds glint on the snow.
I am the sunlight on ripened grain.
I am the gentle Autumn's rain.
When you awaken in the morning's hush,
I am the swift uplifting rush
of quiet birds in circled flight.
I am the soft stars that shine at night.
DO NOT STAND AT MY GRAVE AND CRY.
I AM NOT THERE, I DID NOT DIE.

1979
J. WALKER
1960 - 1993

This truth was written on a small board next to the headstone of a grave at the cemetery in Kernville, California, seen by Bill Huth in December of '96, every instant enlightenment shines brighter on Creation.

BILL HUTH

VOICE OF CREATION

DO BETTER THAN YOUR BEST

What ever you, as a part of the Spirit, take on to do, do with all of your might; for through the word of death you shall be reborn in the Spirit which is the heart of Life and which continues to expose a less complicated way of existence. Idleness likens a man to another who folds his arms and devours his own flesh. Polish your Soul; God is your partner. When you approach another understand this; you are both in the hands of the Creator who through both bodies has been there, as you, since the beginning. To yearn for an existence in Heaven can be compared to a person standing on a mountain of gold searching for a penny. Holding an ill feeling for another likens itself to a machine with no neutral.

Neither more or less do things happen to you that happen to others; As the spider bites one so does it bite another. Everyone stumbles, falling never has an end. Take advantage of distress; learn by experience. Wisdom has always been over and above brawn; a strong man is one who stands beside the meek; in silence. Confusion is the result of misplacement; together they will bring insecurity.

Exist in darkness as well as light; They both have their advantages. Our brain is fed while our body sleeps; a body that does not sleep soon becomes insipid. Who can tell day from night if they can not see? only the blind; there is really no difference.

BILL HUTH

THE PROGRESSION OF LIFE ON EARTH

If we compare the various groups of animals of the present day we shall find that they can be arranged in a series that gradually leads from the simplest to the most complicated—from the lowest to the highest. The lowest forms are minute specks of jelly-like substance in which feeding does little beyond helping multiplication. Next higher we find animals of more elaborate structure, in which feeding is improved by the presence of small muscles that make grasping easier. Muscles next form a greater proportion of the body and are used for moving about; and in forms still higher we find nerves to control them. Muscles for locomotion work better by being attached to a skeleton, and in the early forms of life this is altogether an outside shell like that of a cockle, a lobster, or a fly. The nerves next gradually become more elaborate and usually tend to be thickest in the head.

New possibilities arise in still higher forms, in which the muscles are fixed to an internal skeleton, around a backbone, and the front end of the nervous system becomes a brain. Next, the blood no longer remains of the same temperature as the surrounding water or air, but is warmed by an improvement in the heart. The brain grows in size and complexity, fostering activity and leading to the development of higher intelligence. Finally, there comes Man, mastering the world by his greatly developed brain.

Beginning by living to eat, the series soon advances toward eating to live. Then comes the reign of flesh, with just enough nerve to make the muscles effective for moving and grasping. Finally, the brain end of the nerve begins to preponderate, so that the animal no longer responds listlessly to its surroundings but improves first in instinct, then in reason, and eventually attains supreme intellectual control.

The question therefore arises whether this regular advance has any meaning. If all animals and Man came into existence at one and the same time in their present forms science might find the meaning of the world of life beyond its ken. The fossilised remains of animals embedded in rocks afford, however, direct evidence that the different kinds appeared on the earth not suddenly, at one time, but in orderly succession, the lower first, the higher later. Existing animals are seen to be merely the scattered and more or less altered survivors of various groups that have had their day one after another during the march of the ages. There seems to have been a slow evolution of life from the lowest to the highest, one group after another flourishing in turn and then dying down, leaving only a few remnants as their posterity. The earth thus records its own history within itself; it writes in imperishable rocks the story of advancing life, and the writing may be as clearly seen and deciphered as the writing on the Rosetta stone, although it is only half a century since Man has systematically attempted to read the story told by the rocks.

The succession of rocks containing fossils was, however, made out in part long before naturalists in general had framed any theories as to the evolution of one group of animals from another, and they therefore were not subject to bias in dealing with the evidence. Indeed, most of the pioneers in geology were firmly convinced that the progenitors of every living thing had been separately and specially created. Fortunately, the inquiry as to the significance of the fossils found began on the western edge of the European continental region, which has in past ages sunk repeatedly beneath the sea and then risen again to become dry land. In thus rising the fossil-bearing beds, which were deposited in successive seas and estuaries, have been somewhat tilted, and their edges have been exposed to view, so that it is easy to examine them and the fossils they contain. More than a century ago William Smith, an English land-surveyor, showed that the order of the rocks that contained these fossils is perfectly clear. Nearly all the chief phases in the succession of life are represented in the old sea beds that now form rocks in the British Isles and the adjacent parts of the European continent. Approximately the same succession has been observed in other parts of the world, and several of the greatest gaps in the geological history of Western Europe have been filled by the discovery of rocks of intervening ages elsewhere.

The order of the formation of the series of fossil-bearing rocks has thus been definitely determined by observing the order in which the layers rest one upon another and by comparing this order in detail in different parts of the world. There is nothing hypothetical in the result of this research. The main difficulty is the imperfection of the record made by the fossils. Generally no part of an animal but its hard skeleton is found; the softer parts have been preserved only in exceptional circumstances. Most of the rocks, at least those of the earlier periods, were formed in seas or estuaries, and so yield remains of land animals only where these have been carried into the water. A fossil is buried by accident and is discovered by accident. We may say that our knowledge of fossils depends on a chapter of accidents. It is not remarkable, therefore, that only a few chapters of the past history of life have been clearly read and that a fact of general significance may be known by not more than a single observation. Nearly every fresh exploration adds something new and shows how much depends on local conditions. So far as it goes, all the evidence points in the same direction—*to the slow and regular advance of the world of life in the way already stated. No conflicting evidence has thus far been discovered.*

The beginnings of life will probably never be known, for there is reason to believe that the earliest animals were soft-bodied, without skeletons. They probably originated in the open sea and acquired hard parts only when they settled down within reach of the surf. By the time that any of them had gained enough skeleton to be regularly fossilized, toward the dawn of the Cambrian period, members of most of their early predecessors had disappeared, so that their earliest history is unknown. Swarms of other soft-bodied animals were living at that time, for more or less vague impressions of them occur in a peculiar bed of greasy shale of the Cambrian period in the Rocky Mountains of Canada.

It is clear, however, that before backboned animals appeared or before animals acquired skeletons, the backboneless groups flourished widely and were at some times and places represented by larger animals than any of their kind of later date. Great armoured cuttlefishes, for example, and gigantic lobster-shaped animals were the rulers of the seas before the earliest backboned animals—the fishes—began to flourish. Soon after the appearance of fishes the lower groups just mentioned lost their leading place, and most of them died out. A new era had begun, in which fishes increased both in numbers and in size. The Old Red Sandstone, both of Europe and of North America, laid down as sand millions of years ago, tells of ages when some of the fishes were stranded in pools that at times dried up. Under these circumstances some of them passed from gill-breathing to lung-breathing animals and acquired paddle-like legs suitable for scrambling about on land.

Thus arose the first backboned animals that spent part of their life on land and part in water—the amphibians, which are now represented by the newts, salamanders, frogs, and toads. Then, through more tribulation of drought and desert in the Permian epoch, there came the equally cold-blooded reptiles--lizards, crocodiles, alligators-- animal capable of living all their life on land. They found conditions so easy that they literally swarmed over all lands and even invaded the air as flyers and the sea as swimmers. They increased immensely in bodily bulk until some of their latest representatives in the Cretaceous period (the period of the Chalk) were the biggest masses of flesh that ever lived on land.

A few of the more progressive of these reptiles rather early began to show signs of becoming something better, and by the time the giants of the group were worn out, the progressives had become warm-blooded animals, with an improving brain and very active legs. These were the mammals, quadrupeds which soon began to suckle their young and care for them in their youth. At the beginning of the Tertiary period they took the place of the giant reptiles, which had disappeared. Birds also took possession of the air.

During the Tertiary period the mammals occupied every sphere of life on the land, and as they became more completely adapted to their surroundings their brains grew relatively larger and more useful. As the flesh-eaters advanced in power of jaw and in cunning, and as the vegetable-feeders gradually acquired teeth for grinding hard grasses and nimble feet for running rapidly on plains, their brains kept pace with their needs. Some of the mixed feeders, which lived in the forest and underwent only slight bodily changes to adapt them for swinging about in trees and to feeding on fruits and small animals, became even better equipped with brain. These were the monkeys and the apes. In the apes the brain was especially complicated, and there is reason to believe that in a few that eventually took to life on the ground the brain gradually became very large. Thus arose the distant ancestors of Man, who is shown by fossils to have existed only in a very late geological period. Man himself, indeed, did not appear until the latest geological period—until many of the other mammals were ready for his use for food and domestication.

Fossils do more than prove this general progression of life on earth. They show that there are definite changes—some of them progressive—in each group as it is traced through successive geological periods. They also show that these changes are more or less gradual, not sudden. Fishes may be considered a good example. The oldest fairly well-known fish-like animals, those of the Silurian period, have no hard parts beyond scales and plates in the skin. We can infer from certain markings in the fossils that they had, inside, the beginnings of a backbone and also of a skull, which contained a brain like that of a fish, but neither backbone nor skull was hard enough for fossilisation. Some of these earliest fishes took to life on the bottom of shallow waters, and their skin-armour thickened into bony plates for protection.

In the next period (the Devonian or Old Red Sandstone) the swimmers as well as the bottom-dwellers gradually acquired an elaborate skin armour that was covered with shining enamel, and hence they are described as "ganoid," from the Greek work for resplendent. Their arm fins and leg fins were stiff paddles, used more for crawling on the bottom than for propulsion or balancing in swimming. Some that had powerful jaws, such as *Dinichthys* ("terrible fish"), became gigantic, having heads three or four feet across and armour in places three or four inches thick. Like our early steam battleships, they specialised in weight of armour, and like these battleships they were soon superseded by rivals which depended for success on swift movement rather than on stolid defence. In a few of the Devonian ganoids the two pairs of paddles were replaced by ordinary flexible fish fins, which were strengthened by fin rays, and in some the paired fins were halfway between these two patterns.

During the next period (the Carboniferous) ganoids with flexible paired fins predominated, but they were still handicapped as swimmers by the low degree of hardening of the internal bones, by the incompleteness of the tail as a swimming apparatus, and by the unfinished mechanism of the fins along the middle of the body above and below. The tail was formed by the tapering end of the body, turned upward to make an upper lobe; the real fin was below this, as in the sharks and sturgeons of the present day.

During the Permian and Triassic periods, which followed, the tail in the more progressive fishes lost its upper body lobe by shrinkage and became a most efficient tail fin, and the middle fins were gradually brought up to the most efficient form. The internal skeleton was also gradually hardened.

Early in the next period (the Jurassic) the backbone in some fishes was completed. Each joint or vertebra was deeply hollowed at each end to admit soft, elastic substance and so to give the great flexibility that is needed for rapid swimming. The bony skull was also completed. At the same time the scales became thinner and deeply overlapping and contained very little bone substance and enamel.

In the following Cretaceous seas there were a few fishes that had the bony support of the tail as well formed as that of most existing bony fishes; indeed, the reign of the modern thin-scaled bony fishes, completely adapted for rapid movement in water, had begun, and the only subsequent changes were those which have given almost endless variety to this thoroughly efficient race. It is also interesting to note that the fishes which achieved these latest developments include nearly all those that are used as food by man today.

To summarise briefly: The first fishes were encumbered with outside armour and their fins were not very well formed for swimming. Next, the paired fins became thoroughly adapted for balancing, and then the tail fin was improved until it became a perfect propeller. After this the inside skeleton became bony and gradually grew more efficient and more complicated, and the scales of many forms became thin. Thus, by progressive stages, in a definite order, fishes were continually improved for locomotion and for feeding in water, and there arose the possibility of the infinite variety found among the existing bony fishes.

A student of fossils recognises that *when any kind of animal shows a tendency to change in some particular part, the degree of this change increases in successive generations, especially if the change at first gives it some advantage.* Among the later ganoid fishes of the Jurassic period there are some that tend to assume the form of a swordfish, which has a powerful tail that fits it for darting as well as for swimming. The snout begins to thrust itself forward at the front of the upper jaw. Toward the later part of the Jurassic period the snout even forms a pointed weapon. In the middle part of the Cretaceous period, which followed, the pointed snout in some of these fishes became longer. Toward the end of the Cretaceous period (represented by the Chalk), the snout is much elongated and occasionally forms even a sharp blade, as deadly as that of some existing sword fishes. The increasing power of the snout was thus acquired by gradual growth, which can be followed in the fossils stage by stage.

Among land mammals, or quadrupeds, the deer are very interesting for the same reason. Fossils show that the earliest deer had no horns, or antlers. The next deer had small antlers, but none of them were forked more than once. A little later, in the Tertiary period, some of the deer had antlers with from two to four prongs. In the later part of the Pliocene epoch, in the Tertiary period, some of the deer, when full-grown, had antlers even larger and more complex than any deer existing at the present day. Indeed, it is probable that these deer were handicapped by their overgrown antlers and so died out.

Overgrowth of a part that has begun to show progressive enlargement is often observed among fossils. The gigantic tusks of the elephants that lived in late Pliocene and Pleistocene times are further examples. Also the great canine teeth of the sabre-toothed tigers, which lived with them. In both these animals the enlargement was doubtless a hindrance and eventually helped to put an end to them. Excessive enlargement of this kind must have been usually a hindrance, but there is one great enlargement, already mentioned, which proved to be an advantage—that of the brain in mammals. The great growth of the brain which led to the appearance of man, with his superior mental equipment, was the natural result of the progressive development of the brain in the higher mammals during the Tertiary epoch. Though the fossil apes were very different from modern apes, they must be regarded as the ancestors of both the modern apes and man. Not all the stages between the ape and man have yet been found, because the higher the brain power the more wary the animals would become in avoiding accidents by which their remains could be buried in the earth; but the few fragments that are known show that the links certainly existed. The teeth and jaws of fossil apes suggest that they belonged to animals which may have been ancestral to man as well as to modern apes, and the oldest known fossil human skulls and jaws exhibit more ape characters than any human skull and jaw of the present day.

The oldest jaws of apes thus far discovered are from the early Tertiary (Oligocene) deposits of Egypt and belong to animals smaller even than the existing gibbons, the smallest living apes. They have a short, bony chin and small canine teeth. By a very slight reduction of the canine teeth and equally slight changes in the molar teeth the heads of these apes would approach in form the modern human head. By an enlargement of the canine teeth and a lengthening of the bony chin they would acquire the jaw of an existing ape.

The jaws of the next higher apes, from the middle Tertiary (Upper Miocene and Lower Pliocene) of Europe, represent larger animals, equalling in size a modern chimpanzee. The so-called "forest-ape" (*Dryopithecus*) now has powerful canine teeth, and so is approaching the modern apes rather than man; but its molar teeth are remarkably human in appearance, and the short, bony chin was less prominent than that of the chimpanzee and gorilla.

Teeth and fragments of jaws of several other apes from rocks of the same age in India show that in this region there must have been more variety among apes than is seen anywhere at the present day. There is, in fact, good reason for supposing that these animals may have included some of the actual ape-like ancestors of man. None of them were larger than a chimpanzee or small gorilla.

The fossil we call *Pithecanthropus* ("ape-man"), from a late Tertiary deposit at Trinil, in Java, is distinctly larger—as large as an average man. It is known by the top of a skull, some molar teeth, and a long, straight thigh-bone. The skull is shaped like that of a gibbon, having immense bony brow ridges, but it is nearly large enough to have contained a human brain, and an impression of the brain cavity shows that it had a few human characteristics. Some authorities, indeed, regard *Pithecanthropus* as an overgrown gibbon; others believe that it belongs to the same family as man. Better specimens are needed to determine exactly its relationships.

The earliest undoubted men are known only by remains of skeletons from Europe, which show some peculiarities of apes. *Eoanthropus* (the "dawn-man") is represented by parts of a skull and lower jaw from a river deposit at Piltdown, Sussex, and is especially interesting as approaching an ape in the shape of its lower jaw and front teeth. It has as good a forehead as any modern man, and the size of the brain case is well above that of the lowest existing savages; but the skull lacks the beautiful dome-shape of the ordinary modern human skull, and the neck must have been unusually thick. The shape of the bony chin is unlike that of man and is almost identical with that of a young chimpanzee. Indeed, the whole of the bone of the lower jaw is remarkably ape-like, and it is shown to be human only by two of the molar teeth, which remain in their sockets. The canine teeth are much larger than those of modern man, and the canine of the lower jaw interlocks with its opposing tooth, as in the apes. The only other known human skull that apparently makes some approach to the same form is a fossil skull of an Australian native found in a river deposit at Talgai, in Queensland.

Another fossil human jaw, found in a sand pit at Mauer, near Heidelberg, Germany, is ape-like in the downward and backward slope of the bony chin. In other respects, however, it is typically human, though it is unusually thick and heavy. It is probably almost or quite as old as the Piltdown jaw, just mentioned. At the beginning of the Pleistocene epoch, therefore, there existed in Europe more than one race of men that resembled apes in the peculiarities of their jaws.

Later deposits in Germany, Belgium, and France—even some so far away as Palestine—have yielded remains of men with a large brain case and typically human jaws, but with the bony forehead inflated into great brow ridges like those of a chimpanzee. These early men are known by almost complete skeletons, because they had learned to bury their dead, and several of their burial places in caves and rock-shelters have been discovered. They represent the Neanderthal or Mousterian man, so called because the first skeleton to attract attention was found in a cave in the Neanderthal near Düsseldorf and the stone tools which this kind of man made were first studied in the cave of Le Moustier, in the Dordogne. Neanderthal man walked with a shuffling gait, not quite upright, as proved by his gorilla-like neck and thigh bone. Indeed, he combined in one body more ape characters than are seen in any other low kind of Man.

The cave-floor deposits and others later than those containing Neanderthal man yield no remains of any but typical modern man, *Homo sapiens.* Some of these remains suggest that the human races of the northern hemisphere were at first less distinctly separated than they are at the present day; but the skeletons found are still too few to warrant definite conclusions. Fossil skulls from Wadjak, in Java, and from northern Rhodesia, in South Africa, seem to show that the lowest modern race that survives in the Australian region was formerly more widely spread in the southern hemisphere. One skull that appears to have belonged to a member of this race, found in a cave at the Broken Hill mine, in northern Rhodesia, is remarkable as having the largest bony brow ridges ever seen in a human skull. It is probably an example of "reversion" to a form common in some ancestral race.

The succession of fragments of apes and men already found among fossils therefore justifies the expectation that further discoveries will reveal a multitude of links between the lower (or animal) and the higher (or human) group. The chain of life is undoubtedly complete to its uppermost limit.

SIR ARTHUR SMITH WOODWARD

ALIGATOR TEARS

I've heard people who proclaim themselves to be able to state opinions that are worldly or matter of fact in their own minds. They say, "only mankind are capable of crying." "They are the only ones in creation who have the ability to shed tears." It has been my experience to see animals in the wild wait patiently at the spot where they last saw their young who were either shot by a hunter, or devoured by another carnivorous species. They will not leave this place for several days, sometimes even weeks awaiting for their young to return. To cry is interpreted in the dictionary; to sob or shed tears; To shout or call loudly. To utter a characteristic sound or call. Used of an animal. When confronted with this, these persons will almost always reply, "I mean they can't shed tears"' If this were true a mote would only be seen in the eye of humankind. Severe stress causes the brain to become faint and this pressure is relieved by excretion of fluid through the sinus and draining from the eyes, nose and mouth. Crying means very much more than sobbing. My own opinion would be that mankind are the only creatures capable of feigning the act of crying. I'm quite sure that I have witnessed many tragedies which involved friends or family members who were undoubtedly crying yet their faces were expressionless, and without the slightest sign of a tear or the sound of a sniffle.

BILL HUTH

THE SPIRIT

EPHESIANS 2:18 For through Jesus we both have access by one Spirit unto the Father.

ACTS 17:28 For in Him we live, and move, and have our Being.

ROMANS 10:12 For there is no difference between the Jew and the Greek: for the same Lord over all is rich unto all that call upon him.

CORINTHIANS 2:10 But God hath revealed them unto us by His Spirit: for the Spirit searcheth all things, yea, the deep things of God.

CORINTHIANS 3:9 For we are laborers together with God: ye are God's husbandry, ye are God's building.

CORINTHIANS 3:16 Know ye not that ye are the temple of God, and that the Spirit of God dwelleth in you?

II CORINTHIANS 6:16 And what agreement hath the temple of God with idols? for ye are the temple of the living God; as God hath said, I will dwell in them, and walk with them, and they shall be my people.

GALATIONS 3:26 For ye are the children of God by faith in Christ Jesus.

GALATIONS 3:28 There is neither Jew nor Greek, there is neither bond nor free, there is neither male nor female: for ye are all one in Christ Jesus.

GALATIONS 5:14 For all the law is fulfilled in one word, even in this. Thou shalt love they neighbor as thyself.

GALATIONS 5:22 But the fruit of the Spirit is love, joy, peace, longsuffering, gentleness, goodness, faith, meekness, temperance: against such there is no law.

BEFORE THE BEGINNING?

In the beginning He created heaven and earth. The earth was without form, without it's own form or without the forms of life, It doesn't explain. And the Spirit of God moved upon the waters. So now we have God, His Spirit, The Universe, the waters above heaven, heaven, the waters below heaven on earth and earth without form and void, I mean silent!

He didn't work too hard the first day; He just gave us light and darkness.

The second day God gave us what we might, or do call Heaven, our most valued asset! So far.

On the third day He created the seas; land and vegetation appear. And He saw everything was good. Genesis I:ll: And God said, let the earth bring forth grass the herb yielding seed and the fruit tree yielding fruit after <u>his</u> kind. "After his kind." In other words, he, the tree, was referred to as "his kind". Naturally there were no discrimination laws at that time and almost everything was referred to as he, him or his; all but mother earth - she hatched it all out of the earth and humankind gave her full credit for that, but that was it, no more. From now on you can forget her; till women's lib.

The main point of the third day was, God referred to the fruit tree as "his", "his" is a living, moving growing part of His, God's, creation. A part of the future... He isn't just a fruit tree dropping fruit, "he" is a living being just as Adam will soon be. not capable of walking but yet a living thinking part of life... And God went on in verse 12 referring to the grass with "his" seed.

I'll drop Genesis at the end of the third day to explain to you that one of the most important parts of a person's belief in evolution is that he or she or "it" must consider themselves through the Spirit of God, or whatever you want to call Him, that God and humankind are a part of all creation: not more or less but an equal part with all creation - creation being all you see, all you hear and all you can't see and all you can't hear. There are n "missing links". Humankind fit in perfectly with the fruit tree, the grass, the micro-world, all animals, reptiles, everything; man and woman, male and female, positive and negative, even fish and snails. Everything is blessed with eternal life. Understanding this is the problem!

When someone speaks of death, they refer to it as the end. This is not true. Death is a word that describes the Spirit, that lives in all creation as an Ether Being that can not be described or explained by any description, which transforms Itself as a part of rebirth in all existence. The primary purpose being rebirth.

The Spirit has always been and forever will be. Our Spirit is Gods Spirit. The benefits all creation receive for putting up with life is that life continually gets easier. Humankind not only take care of each other but now we have hospitals for almost any of creations members: even a psychiatrist for dogs, flowers or anything else. Living has always been fun, but it has also been a lot of work and problems; but it has always been worth it.

Soon we will understand; We are born into creation from the Spirit. We live life in The Soul as partners with this Spirit called God. Through what we call death we are then reborn into the Spirit; God. We live and we die and are born; At all times we remain an integral part, or better than that a Partner, with God.

PERHAPS REINCARNATION IS THE AUTHORITY.

BILL HUTH

COYOTES EAT GRAPES - CATS EAT COOKIES

All of my life, or at least seventy years of it I have not been able to understand why the human race has refused to accept any creature other than humankind to be their equal. We can't fly. We can't stay under water and swim like a fish. We are unable to hibernate in either snow, sand, mud or clay; there are many things that mankind are unable to do just as other creatures are unable to do many of the things that mankind can do. Again there are many of God's creation that "marry for life." Even when one of the pair is separated by any means the other will remain "true" in hopes of in someway being Spiritually reunited. Even without enjoying the benefits of mental and physical evolution there are those who decline the relationship of other creatures. To inform another that you feel the tree to be a part of your roots or essence is almost or practically out of the question. Some persons bend a little and call themselves vegetarians, eating only vegetables, maybe a little fish.

Not changing the subject; it is very easy to take off from this thought of not killing an animal to eat it but what is the difference of "killing" a cow or a tomatoe; (I like to spell it this way, it seems like it adds to the personality of the tomato) there is none! There are many plants which are carnivorous. In the oceans even more plant like species gobble up small fish and urchins. Gorillas live on vegetation. Bears like all sorts of berries and fruits. Even Coyotes eat grapes and vegetable... When I was a child we fed our cats liver; I never in all my life thought I'd see a cat eat a cat cookie.

About fifty years ago when Maxine, my wife and I were first married she told me about how her dad used to put sauerkraut in front of their cat and then squeeze his tail and he would eat it. I've seen elk, who have just been shot at, under nervous tension run right out in the middle of a meadow and start eating. Many people have told me that when they get nervous or excited they will run to the refrigerator.

Getting back to the subject, killing is not really killing. The Spirit is our Savior; better yet, the Savior of all creation. Everything is converted to energy of some sort: obesity or taking more than your share is one problem and the other is wasting food or fuel. Obesity is some cases is storing for the future. Eating and eating until the fat kills you, that is a deadly problem. TEN 4.

BILL HUTH

THE EVOLUTION OF PLANTS

In the conservatories of the Brooklyn Botanic Garden there is an exhibit designed to give a bird's-eye view of the plant kingdom.[1] The specimens are arranged on a bench in the form of a tree, with a trunk and lateral branches (Fig. 1). The trunk represents the main course of plant life through the ages; the branches are the great groups of plants. The plants now living on the earth are to be thought of as representing the tips of the branches of the genealogical tree.

Near the base of the trunk, on the lowest branch, are specimens of some of the simplest plants known. As we pass from these toward the other end of the bench we find plants of gradually increasing complexity, until we come to the orchids and composites at the topmost twigs.

Along the trunk of this family tree is a label indicating the changes met in a series of plants arranged in this order. The points where the branches leave the main trunk are "mile posts" calling attention to definite changes there represented. This long label is here reproduced, with the "mile-posts" in heavy-faced type. The names of the great groups of plants are in large and small capital letters:

[1] The exhibit here described was planned and installed by Dr. Alfred Gundersen, Curator of Plants, Brooklyn Botanic Garden.

ALGAE.

Nearly all grow in water, with spores unprotected.

From water to land

LIVERWORTS AND MOSSES.

Have spores in a protecting spore case and grow on moist land. Have no true roots but have root-like organs (*rhizoids*).

From rhizoids to roots

CLUB MOSSES.

Have roots and water-conducting tissue and grow on drier land. Leaves small. Reproduce by spores. One spore case. Each spore develops into a small sexual plant (*prothallus*).

From small leaves to large leaves

FERNS.

Spores usually all of one size and germinate on moist ground. No seeds.

From spores to seeds

CYCADS.

Two sizes of spores, which grow to small sexual plants that develop on the parent plants, thus producing seeds. Cycads have swimming sperms, as do all the preceding forms.

CONIFERS (cone-bearing plants).

These and the flowering plants do not have swimming sperms. Cycads and conifers have *naked seeds*—that is, they are *gymnosperms*.

From cones to flowers

ANTHOPHYTA (flowering plants).

Seeds enclosed in a covering—the fruit. Some seeds have two seed leaves (*cotyledons*), some only one, thus forming the following groups:

Dicotyledons.

Two seed leaves; leaves netted veined; parts of the flower in fives or fours.

Monocotyledons.

One seed leaf; leaves usually parallel veined; parts of the flowers in threes.

The above summary takes account of all the great groups shown in the exhibit except the *fungi,* which form one of the lower branches, just a little above the algae. The fungi resemble the algae in essential characters, except that none of them has the green coloring matter of plants, known as leaf-green or *chlorophyll.* They appear to be algae that have permanently lost this substance, but we do not yet know enough to speak with assurance as to the origin of the fungi.

In contemplating this exhibit, or the plant kingdom itself, here represented in miniature or by samples, one is impressed

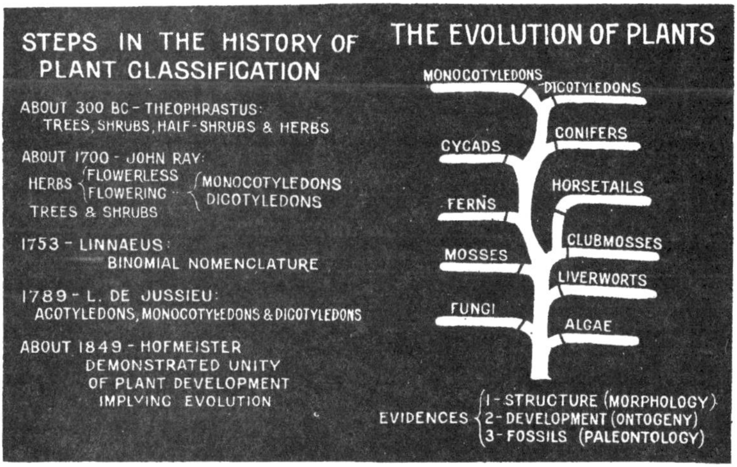

FIG. 1.—At the left, a brief outline of the history of plant classification. At the right, a genealogical tree indicating that the modern groups of plants were descended by evolution, not one from the other, but from preëxistent ancestors, which were also genetically related. This conclusion is based upon a study of the structure (*morphology*) of existing plants, the life-histories of individual plants (*ontogeny*), and a comparative study of the morphology of fossil plants (*palaeontology* or *palaeobotany*) and modern plants (*botany*). (*From the label for the Evolution Group at the Brooklyn Botanic Garden.*)

by the fact that, amid the endless diversity of plant forms, it is possible to bring order out of apparent chaos. Men have been trying to do this with plants for more than two thousand years. What a long, hard struggle it has been to try to understand Nature! About 300 B.C. Theophrastus, a

Greek botanist, a pupil of Aristotle, who was also a botanist, classified plants, according to their most obvious resemblances and differences, as trees, shrubs, half-shrubs, and herbs. This was a very superficial classification, but it took centuries of study by many keen minds to enable us to distinguish between essential and superficial or accidental differences and likenesses. According to the classification of Theophrastus, roses and apples fell into quite diverse groups, but in the modern classification they are placed in the same group. Other systems of classification are briefly indicated in Fig. 1.

Again, in contemplating this exhibit, one cannot help but ask himself the question, "How did all this orderly diversity come about? Have all of these various kinds of plants always existed? If not, which existed first? If they have not always existed, by what method were they created?"

It has been very natural for men to overlook the last question, and merely inquire, "By *whom* were they created?" This is a very proper question to ask, and full of absorbing interest, but if one has the scientific type of mind, he is not satisfied with this question, nor with any answer that may be given to it. We said above that one cannot help but ask himself the question, "How did all this orderly diversity come about?" But the scientist does not ask *himself* this question; he puts the question directly *to Nature* and seeks his answer there. He wishes to know not only who, but *how*.

Each question is important, but the answers are likely to lead in different directions. One who was content merely to know who made the first telephone could never have invented nor helped to invent the radio; that could have been done only by one who insisted on knowing the how and the why—the structure, the mode of action, the underlying principles of the Bell telephone. In acquiring this knowledge it was not necessary for him to forget the inventor of the telephone, to deny his existence, nor to cease to admire him and his work. Moreover, by understanding the telephone he was in a position to understand its inventor more intelligently and to regard him with more admiration and reverence.

Do we enjoy the modern delicious varieties of fruits and vegetables and the wonderfully beautiful horticultural forms and colors of flowers? These were produced, not by Nature unaided, but by Nature aided by Man. The success of men in breeding the numerous horticultural varieties of plants depended upon their understanding the processes by which new forms of plant life come into existence.

But an understanding of the *method* by which the present condition of the plant kingdom was brought about has value from an entirely different point of view. It gives us a more intelligent comprehension of Nature, it widens our intellectual horizon, it reveals a world of law and order, not of caprice and chance, and it enables us better to understand ourselves and our relation to the world in which we live.

In the earlier periods of intellectual inquiry men endeavored to reach an understanding of Nature by philosophical speculation. But, as Mackenzie [1] has well said, just as the special sciences cannot furnish us with those ultimate explanations for which the human mind inevitably looks, so "no purely philosophical speculation can tell us about the particular structure of the world in which we find ourselves." There is only one source of evidence and light, and that is the study of Nature itself. If we would know how the present condition of the plant world came about we must study plants. Let us, then, briefly present some of the more important general truths that have been brought to light by the study of plants.

[1] Mackenzie, J. F., *Elements of Constructive Philosophy*, p. 308.

1. *Biogenesis.* In the first place, studies of great thoroughness and accuracy have led biologists to reach the unanimous conclusion that every living thing comes into existence as the offspring of other living things somewhat similar to itself. There is no other method known by which living things *now* come into existence. This principle has been tersely stated by the familiar Latin motto, *Omne vivum e vivo* (all life from life).

To be sure, there is the ultimate question, How did the first living organisms come into existence? There has been much speculation on this question, some of it based upon painstaking experiment. From what we know of the geological history of the earth we are forced to visualize an early condition when life in any of the forms now known could not have existed. If that is true, then there must have been a time when living matter first came into existence and, of course, from non-living matter. At the Richmond meeting of the American Chemical Society in April, 1927, Dr. Victor C. Vaughan, discussing a chemical theory of the origin of species, noted that although no chemist has yet awakened dead matter into life, chemists have learned how to synthesize, out of inorganic matter, substances formerly found only in plants and animals. Calling attention to the recent discovery of particles smaller than bacteria that pass through a porcelain filter and grow and reproduce like living organisms, Dr. Vaughan contends that the lowest forms of life have come into existence by chemical processes.

Our present inquiry, however, concerns, not the origin of life, but the *method* by which the present condition of the plant world has been reached, *granted the existence of living organisms to start with.* However diverse existing organisms may be, the principle of biogenesis compels us to conclude that they have come into existence by descent from preëxisting organisms.

2. *Gradual change.* Either conditions have always been as they now are, or else there has been a change throughout the vast aeons of geological time. The evidence on this subject consists of fossils found in the rocks of the earth's crust. The more recently formed rocks (such as those of the Tertiary period) contain fossil remains of plants very similar to those now living—in fact species of the same genera. But as we go down the geological scale to older and older rocks all evidence of species now living gradually disappears. Moreover, we find abundant evidence in the older rocks of the existence of forms not represented in the fossils of the more recent rocks and not found at all in the vegetation of to-day. Obviously, there has been a profound change in the vegetation of the earth. Forms appear, persist for awhile, and then die out, giving place to new forms. Moreover, the geological and biological evidence forces us to the conclusion that *this change has come about gradually.*

3. *Evolution.* By studying the comparative anatomy of forms in successive geological periods we learn that they resemble one another just as they would if they were related to one another like parents and offspring. When we contemplate this fact in the light of the principle of biogenesis the only logical conclusion we can reach is that the plants of one geological period have been derived from those of a preceding period by a process of descent with gradual modification. This is what is meant by *organic evolution.*

4. *Hypothesis, theory, fact.* When men study the phenomena of nature they get ideas suggesting explanations of what they observe. These ideas are of the nature of guesses, but they are rational guesses, which are fully warranted by the contemplation of the facts observed. Such a

logical guess is called a *hypothesis,* from a Greek work that means supposition or "guess". The next step, of course, is to try to prove the correctness or incorrectness of the hypothesis. This may be done by reasoning out what consequences ought to follow *if* the hypothesis is correct, and then making further investigation to ascertain whether such consequences do follow in reality. If the hypothesis is found untenable it is abandoned. As Huxley once pointed out, the pathway of the history of science is strewn with abandoned hypotheses.

If a hypothesis is found to be valid, if the consequences postulated are realized in fact, then the hypothesis comes to be called a *theory.* A theory is a hypothesis that has been found to be in harmony with all the known facts, or with a vast majority of those facts. Eventually the conception called a theory may no longer be regarded as a theory but may be considered an actual *fact* or *truth.* Thus, our conception of matter as composed of atoms was at first a hypothesis; all that we knew suggested that idea. After undergoing rigid tests the atomic hypothesis became the atomic theory. Now we have, so to speak, handled atoms, and separated them into their component parts, so that atoms are no longer regarded as hypothetical or theoretical things but as actual facts.

So with the conception of the evolution of living things from earlier, simpler organisms to those now living. The idea was first a hypothesis, then a theory; and probably no living student of plants now doubts that evolution is a fact —that is, he believes that the present condition of the kingdoms of plants and of animals was attained by the process of evolution.

In the evolution of the plant kingdom the evidence available forces us to conclude that the earliest organisms were protein compounds endowed with these peculiar attributes:

1. Ability to take in matter from without and transform it into matter like themselves. This we call *metabolism* and *nutrition.*

2. Ability to *grow*—to increase in size and weight.

3. Ability to *reproduce* their kind.

4. Ability to *detect changes* in their surroundings and to *react* or readjust themselves to the changed conditions. This we speak of as ability to detect and to *respond to stimuli.*

It is also probable that the earliest forms of life were so simple that they could be regarded as neither plants nor animals, but merely as *organisms* or "living things." Such organisms are well known today. The *viruses,* which cause the so-called virus diseases of plants, are possibly of this nature. They behave like living things, but they are so small that they cannot be seen with the most powerful microscope. This means that their greatest dimension is less than one-half the wave-length of light.

One group of organisms, known as slime molds (Myxomycetes, Fig. 2), at one stage of their existence so closely resemble the tiny animals (animalcules) known as *Amoebae* (singular, *Amoeba*) that they can hardly be told apart. Both are naked bits of protoplasm, capable of motion and locomotion. Zoölogists have regarded them as animals; botanists have contended that they are plants.

The similarity between animals and plants in their essential life processes has long been recognized by biologists and was forcibly presented by Claude Bernard in his classical *Leçons sur les Phénomènes de la Vie Communs aux Animaux et aux Végétaux* (1878-79). In the processes of respiration, digestion, cell-division, growth, reproduction, transmission of heritable characters, the possession of irritability and the power to detect and respond to stimuli, and in other physiological processes, they are essentially alike.

Early in the development of living things one group of plants acquired the ability to manufacture that wonderful substance, *chlorophyll,* which gives the green colour to all foliage, and these primitive chlorophyll-bearing organisms must be regarded as the ancestors of all plant life.

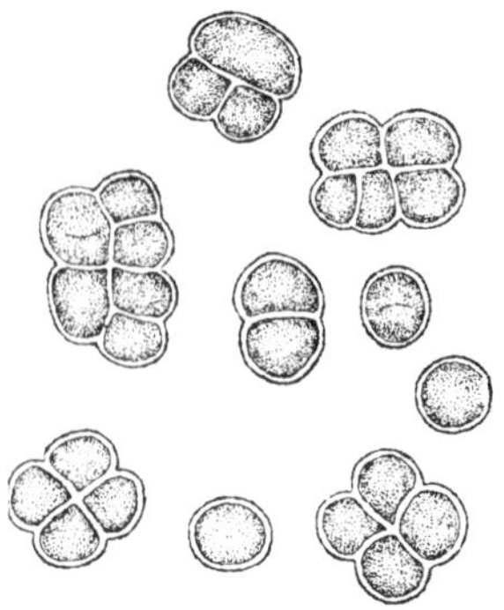

FIG. 3. — Individual plants of a simple green alga (*Pleurococcus vulgaris*), showing reproduction by cell division. The cells tend to remain attached after dividing, thus forming a transition from a unicellular to a multicellular plant.

The simplest chlorophyll-bearing plants to-day are the unicellular green algae, such, for example, as *Pleurococcus* (Fig. 3). These reproduce only by *cell-division.* Other green algae, such as green silk, or *Spirogyra,* reproduce by both cell-division and *cell-fusion.* The introduction of cell-fusion into the life histories of organisms laid the foundation for the development of sex, for cell-fusion is the essential process in sexual reproduction.

The modern representatives of the other great groups of chlorophyll-bearing plants, such as the mosses, ferns, club mosses, little club mosses, and conifers (Figs. 4-7), illustrate definite advances in evolutionary progress, but they do not form a genetical series—that is, they do not bear to each other the relation of ancestor and descendant. Some students incline to the opinion that all the great modern groups of plants have descended from one main hypothetical fern-like branch, the *Primofilices,* which can be traced back to the dawn of the fossil record but is now extinct. From the Primofilices there descended cycad-like forms (*Cycadofilices,* cycad-ferns), also now extinct, but known from abundant fossil remains. They have also been called *Pteridosperms,* or seed-bearing ferns (Fig. 8). From these descended cycad-like plants, which had features resembling those of modern flowering-plants of primitive type and which are called *Pro-angiosperms.* From this stock are descended the modern *cycads* (Figs. 9 and 10) and the two great groups of flowering plants — those with two seed-leaves (dicotyledons —magnolias (Fig. 12), buttercups, roses, bell-flowers, dandelions, etc.), and those with one seed-leaf (*monocotyledons* — lilies (Fig. 14), grasses,

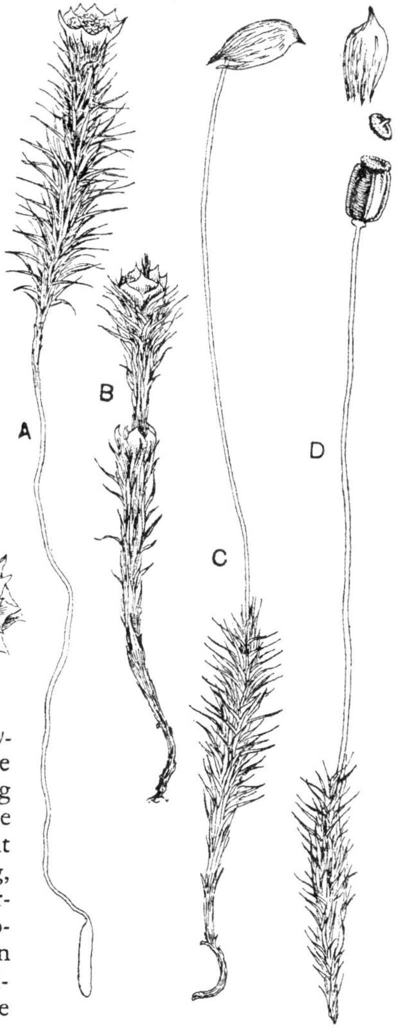

FIG. 4.—Hair-cap moss (*Polytrichum commune*). A, male plant; B, same, reproducing vegetatively, growing from the tip of another; C, female plant bearing a spore-case on a long, slender stalk. This spore-bearing phase of the plant (sporophyte) is developed from an egg-cell after it had been fertilized by a sperm from a male plant.

FIG 8.—A seed-bearing fern (*Neuropteris heterophylla*), one of the Pteridospermae or Cycadofilices. The leaves resemble those of the royal fern (*Osmunda regalis*), and some of those shown bear seeds. Note that the unexpanded leaves are circinately coiled, as in modern true ferns. The two lower left-hand figures show fronds with seeds attached. Restoration by Miss Janet Robertson. (After D. H. Scott, *Extinct Plants and Problems of Evolution,* with the permission of the author and The Macmillan Company, Ltd.).

orchids, etc.). This, in bare outline, is the monophyletic hypothesis (Fig. 11).

Other students, on what they believe to be equally good evidence, postulate two primitive main branches, appearing as distinct at the dawn of the fossil record—the club-moss stock, or *Lycopsida,* and the fern stock, or *Pteropsida,* from which have descended the modern conifers and flowering plants and their ancestors. This is one of the polyphyletic hypotheses (Fig. 13).

The solution of the question of the "family-tree" of plant life may be roughly likened to the task of putting together a picture-puzzle, in which many of the pieces are not understood and some are perhaps temporarily or permanently lost. If we had a museum collection of specimens of all the kinds of plants that have ever lived, botanists believe that such specimens could be so arranged as to represent their genetic relations and to give us a true picture of the evolutionary development of the present plant world. But probably no such collection can ever be made. We are continually finding, with more or less certainty, where this or that piece belongs in the picture, and lost pieces are continually being discovered as fossils in the rocks or as facts disclosed in the laboratory and field.

Again—to use once more the illustration afforded by the picture puzzle—it is the difficulty of the problem that fascinates the scientist, and it is the modicum of his success that lures him on to further research; he finds his reward in the quest and in the satisfaction of making some contribution, however slight, to the ultimate, but probably unattainable success.

It is one thing, however, to accept evolution as a fact and quite another thing to explain the *method of evolution*—how this gradual change or series of changes has been brought

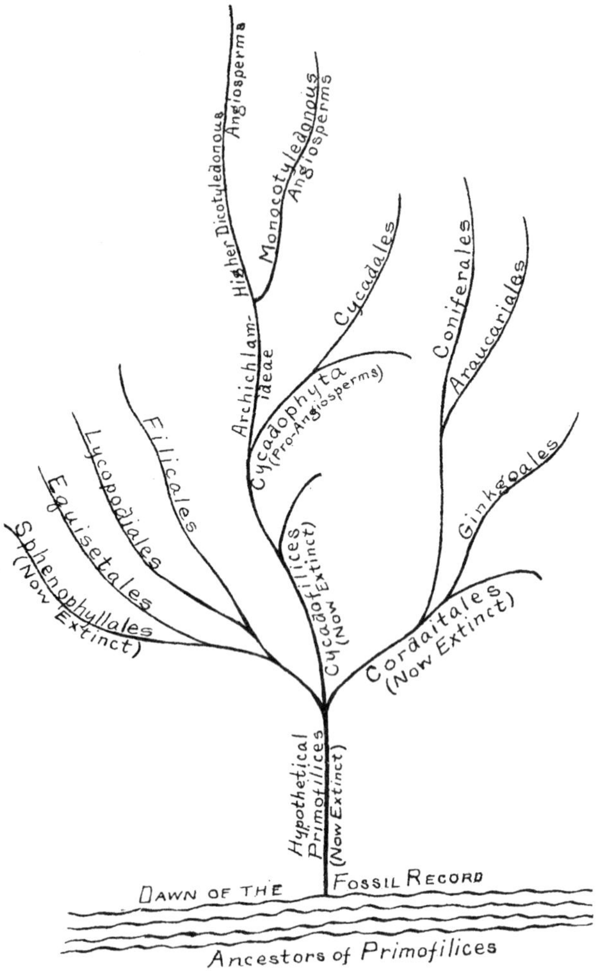

FIG. 11.—Hypothetical genealogical tree showing the ancestral line of the modern plant groups (orders) according to a monophyletic hypothesis. (Compare Fig. 13.)

Reproduced, by permission, from Gager's *Fundamentals of Botany*, published by P. Blakiston's Sons & Co.

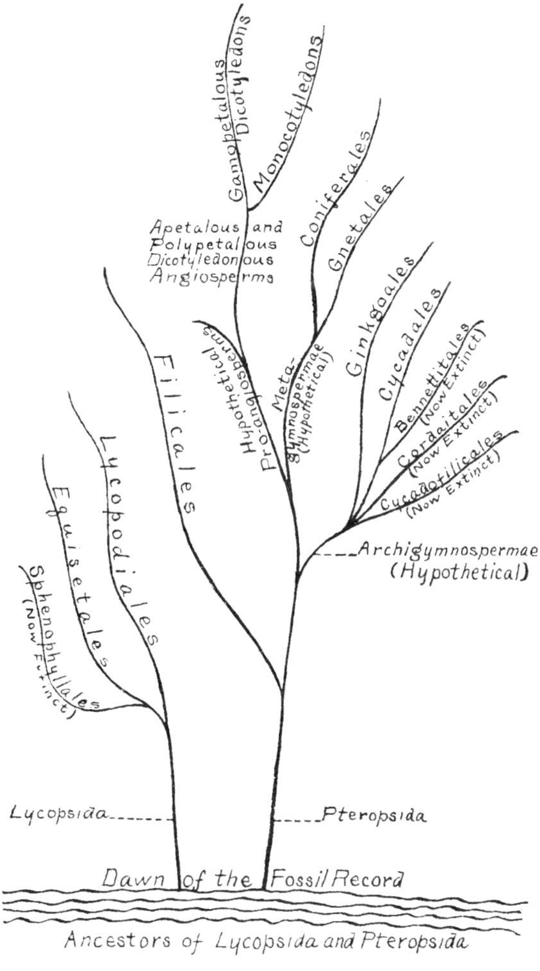

FIG. 13.—Hypothetical genealogical tree, showing the ancestral line of the modern plant orders according to a polyphyletic hypothesis. (Compare Fig. 11.)

about. The theory of Natural Selection, proposed in 1859 by Charles Darwin, is the most fruitful of several that have been proposed. It has recently been discussed so fully and so frequently in daily newspapers and in popular and technical periodicals and books that it need only be mentioned here. That new plant forms may be derived from preëxisting forms by the process of descent with modification has been demonstrated by actual experiments, culminating in the classical work of the Dutch botanist, de Vries. The method by which this may be brought about has been outlined by de Vries in his *mutation theory.* The mechanism of mutation and of inheritance has been worked out in detail by Gregor Mendel and more recent students of genetics, who have extended

FIG. 14.—Turk's cap lily (*Lilium Martagon*). One of the monocotyledons—the group of plants having one seed-leaf or *cotyledon.* There is evidence that the monocotyledons form the most recently evolved group of plants.

Reproduced by permission from Gager's *Fundamentals of Botany.* Blackiston Sons & Co.

and elaborated the pioneer work done on this problem by Mendel.

In summary it may be said that the plant kingdom presents itself to us as a multitude of organisms of various degrees of complexity, ranging from one-celled algae to multi-celled organisms such as orchids and chrysanthemums. The present vegetation of the earth differs profoundly from that of preceding geological ages; and amid all the present and past diversity of form there is evidence that leads to only one conclusion, namely, that the various forms of plant life are genetically related—that the newer and more complex types have been derived by descent (with modification) from older and simpler types. A mechanism has been worked out along lines suggested particularly by Darwin, de Vries, and Mendel, which offers a *partial* but rational explanation as to how these evolutionary changes may have been and probably have been accomplished.

In particular the fact to stress in such problems as those here discussed is that they cannot be solved by philosophical speculation; they can be solved only by first-hand study of plants themselves. In our quest of the elusive thing we call truth, whether in science, religion, politics, or any other department of human thought, the most conspicuous historical feature we note is change, revision, and continued research for new and more reliable information and interpretation. What one generation ties to, the next rejects, but not *in toto*. A residuum remains, which we believe represents the truth. Some progress is made by each generation. The discovery of new facts may necessitate the radical revision or even the abandonment of old ideas, but the only things that should cause us grave concern would be the cessation of the discovery of such facts (for each revision takes us one step nearer to ultimate truth) and the closing of our minds, by prejudice or otherwise, so that we could not entertain new truths nor revise our old conceptions in the light of new and revitalizing evidence. C. STUART GAGER

" SOUL OR SOLE "

In the Old Testament Moses told us God created man. Genesis 2:7; and the Lord God formed man of the dust of the ground, and breathed into his nostrils the breath of life; and man became a living Soul.

Moses didn't say a living Spirit or a person, but, "a living soul." Not a man with a Soul, or a living man with a Soul; just a man with a living Soul.

Webster's New World Thesaurus tells us that a Soul is a disembodied Spirit; a disembodied Spirit means according to Websters Dictionary, a person divest of his Spirit or a person who's Spirit has been taken away from him or her. So we seem to be at a stand-still from the "beginning."

Webster's Dictionary also says, Soul is the immaterial essence, animating principle, or actuating cause of an individual life. The Spiritual principle embodied in human beings, all rational and Spiritual beings of the Universe; I guess this means everything when your an Evolutionist.

Therefore in reality God breathed into the nostrils and created a living Soul! Also Webster's New World Thesaurus says, a sole is only one, no more than one, remaining.

So according to our national or English work or language defining books we are speaking of God creating one Soul or Sole, and then walked over to the east side of Eden and planted a garden; and there he put the Sole whom he had formed; and out of the ground He, God, made every tree to grow that is pleasant to the sight, and good for food; the tree of life also in the midst of the garden, and the tree of knowledge of good and evil. We will talk about these two trees a little later; our chief concern now is Soul and Sole.

" SOUL OR SOLE "

Then Genesis 2:10 continues. If you'll read your bible it says, "And a river went out of Eden to water the garden; and from thence it was parted, and became four heads. The name of the first is Pison: that is it which compasseth the whole land of Havilah, where there is Gold. And I guess this Sole might be able to use a little gold, that never hurt anysole.

We'll continue now, jump down to Genesis 2:19. And out of the ground the Lord God formed every beast of the field, every foul of the air; and brought them to Adam to see what he, Adam, would call them; What I really don't understand, positively, but I've got a good idea, is where did this Sole pick up the name of Adam? It seems to be a little strange; at times though both Moses and his brother and his God seem a little strange.

Personally I have a feeling that neither Moses, his older brother Aron, or their God wanted to be connected with giving Adam the name of Adam. After all none of the first people he created on the sixth day had any names either; only man and woman; maybe this Sole said to Moses God, "just call me Adam." But in my heart I feel that the real truth, all one hundred and ten percent of it was this. Just a little west of Eden was a country called Greece. Here things were a little different and these people were what you might call factualists; today you might call them scientists. They also had a few weird gods too; more important though they had a tiny thing that you could not see, and it seems that the Greecians taught their citizens that everything was made up of these atoms and that you couldn't see them and so Moses and his brother felt like these atoms were a sort of Spirit and they called them Adam too; only they spelt it a little different. Sort of like Sole and Soul. Atom became Adam and everything was cool; nobody to blame for the mistake; even their god was clean, Adam had picked his own name!

Now let's get back to Soul and Sole. Evolutionists

understand that we started almost the same time as Adam; from nothing; but alive! Then we grew into something and were covered by a protective coating called a Sole. We bounced around and abused this layer of protection for hundreds of millions and millions of years until we finally reached a kind of a rough form which we are wearing today.

Naturally with all of the abuse our sole took, the protective coating became tougher and tougher and we began calling this outside layer a sole ourselves. At first we grunted it out and used a lot of body language. We then pointed out to our friends and neighbors in the next cave that what the sound meant was the bottoms of our feet, our elbows, knees, palms, buttox and probably our greatest feature and one we could really be proud of our face; complete with two of our toughest parts, a tongue and a set of lips; these two could stand more abuse and heal faster than any other part of our sole. The intermediate skin made the body a complete Soul as Moses and Aron would soon call it.

As time ticked on the bruised body type sole kept improving and our tongue learned to give our grunts and screams a little character and words turned into sentences and we could now point to our bodies and call them a sole, which rightfully is what we are wrapped up in. Everybody including all creation cattle, fowls of the air, fish of the sea, beasts of the field, trees, plants, reptiles, bugs everything had "It's sole." As a matter of fact our sole became a personal means of identification or better yet recognition for ourselves.

Now we run into Moses and his older brother Aron and their god. god, theirs, declared that Moses was a good front man, but it was Aron who had the silver tongue, and also a way about himself for gaining the confidence of the descendants of Adam and Eve so he put him, Aron, in charge of all the hocus pocus; Moses was still the contact for their god though.

Just to make things easier and make more sense, let's get

back to the sixth day, just before he created Adam and talk about the man and woman he created first; that I feel sure were going to be set up as scapegoats in case something went wrong with Adam and Eve and their descendants.

Well Adam lived to be nine hundred and thirty years old and never had a real problem, so everything went okay up till now with the brothers. At the time Adam lived, a year was one month so in reality Adam passed on when he was around seventy-five years old; our time. By the time their story got this far Moses began asking Aron, "what's up with our people calling themselves soles?" Aron worked on this for a while and told Moses he would have to talk things over with their god to see if there was some way they, members of the tribe of Adam, could get into heaven and retain their soles so they could be recognized by their friend, who would be there too. Moses ran up the mount, talked things over with his god, came back down and told Aron that anysole who wanted to get into heaven with his sole would have to be clean of sin and we're switching the spelling of sole to soul, and making a sort of inner Spirit out of it; You might even call it a Holy Soul.

Adams line of descendants then started. They, Adam and his wife Eve, had Cain and Able; of course the biggest news of the century was when Cain slew Able, but he got a couple of attorneys and beat the case then went on to raise a family himself. Their first son Enoch even built a city and named it Enoch. Then Enoch had Irad, Irad had Mehujael and Mathusael, he had Lamech and Lamech started something new and legal; he went out and got himself in twice as much trouble. He had two wives, Adah and Zillah. All of this family tradition went on till Lamech finally got a son named Noah.

In the meantime their god was losing confidence in life on earth. The Cainites began marrying the Sethites. The Sethites were descendants of Seth who was Cains younger brother. It was really getting messy around Enoch's town of Enoch.

Some of the sons of god saw that the daughters of men (I guess men were the descendants of the man and woman created before Adam) were blonds and they took then as wives, of all which they choose. "Of all which they choose." I wonder just how to interpret this sentence? Well we'll go on anyway. Also at that time there were Giants. Well that's a little different, seems like Moses' god would have told us about that before. Wonder if they came from Adams line or men's line. Sounds like they were Vikings; I guess they got mixed in some way.

Well Moses' god liked Noah, who was born in the chapter before the last one, so he gave him a hot tip to build a boat and the plans for building it just the right size. When he finished it Moses' god told him to load it up with his wife and their three sons and the son's three wives and a pair of everything that Moses' god created in the first five days; and that really made the boat a little over crowded, this would amount to around 20,000 according to todays rules. He, god, didn't say anything about food; I guess it was sort of like Jesus and the fish story. Anyway if they did, it had to be eaten and then passed through and out onto the deck or where ever they were; looks like it should have been a bit deep by the time the cruise ended.

So it came to pass in the six hundredth and first year, in the first month, the first day of the month, evidently this was from the first day of creation and not since the first day of the flood, that the waters dried up and the ark was resting on top of the highest peak on Mount Arat. After a trip like that you'd think Noah would be a little pissed off but he just smiled, cut a hole in the bottom of the boat so everybody wouldn't have to jump and roll down the mountain. After everyone got off the boat including the birds and the bees, Noah and his god sat down and had a little talk. Noah thanked god for letting him start a whole new civilization and god promised him that he would never make it rain like that again and just to prove that he wouldn't he made a big rainbow in the heaven right then and there. "This bow will be a token of a covenant between me and the earth." Evidently he didn't want to trust man anymore so now he was dealing with mother earth, the mother of all creation.

Moses story, Genesis, just gets deeper and deeper; the more or rather the further you read the worse it gets. Every time any of Adams descendants or the descendants of men do anything or make a bad move Noah's god jumps right on Noah's ass and tells him that he had better straighten out his tribe or he's gonna strike them with lightening or make the earth tremble and shake or some other weird thing that will make them tremble and run for their caves. And on top of all that they had to keep bringing him cattle, sheep, doves and anything else they had for a burnt offering so they could clean up their sins so they could get into heaven wearing their sole, I mean soul!

Trying to solve the problem of having Eternal Life is nothing; you only have to understand. Getting out of It is the real punch!

BILL HUTH

WORDS FROM THE BOSS

"There is a grandeur in this view of life, with its several powers, having been originally breathed by the Creator into a few forms or into one; and that, whilst this planet has gone cycling on according to the fixed law of gravity, from so simple a beginning, endless forms most beautiful and most wonderful have been and are being evolved."

CHARLES DARWIN

NEW TOYS FROM OLD IDEAS!

"Nature! We are surrounded and embraced by her; powerless to separate ourselves from her, and powerless to penetrate beyond her.

"She is ever shaping new forms; what is, has never yet been; what has been, comes not again. Everything is new, and yet naught but the old."

GOETHE

CORNER OF DARWIN AND GOETHE

In the early fifties I lived in Mexico City, on the corner of Darwin and Goethe.

BILL HUTH

THE STORY TOLD BY FOSSIL PLANTS

STUDENTS of evolution are at present interested almost exclusively in experimental studies that may disclose its causes, an extremely difficult problem even with the simplest and most rapidly multiplying organisms. Such studies, however, afford the only logical approach to an answer to the question "Why?" The answer to the question "How?" is given best in the geological record. Palaeobotany—the science of the world's oldest plant life—has this advantage over all other methods of finding an answer to this question: the student, to borrow a simile from written history, is dealing with the original documents in so far as they are preserved—the fossil plants—and he finds them in their actual order of succession. Our main task here, then, is simply to tell the story of the procession of the myriad of plant forms across the stage of the past. Before that story is told, however, the general facts and principles illustrated by fossil plants may be very briefly set forth.

First among these is the fact that plants underwent a gradual transformation from simplicity to complexity and were differentiated in both structure and habit in successively higher groups, thus exemplifying the universal principle of evolution. The earliest plants grew in the water, but gradually the main theater of plant operations was transferred to the land.

Each successive group of plants that appeared upon the scene illustrates a second great principle, which is called "adaptive radiation"—that is, from time to time, by progressive modifications, certain groups became dominant, such as the club mosses, horsetails, and seed ferns of the Carbonif-

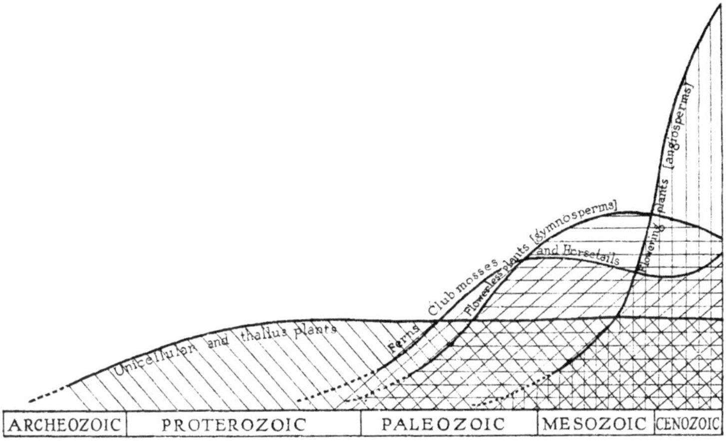

FIG. 1.—Diagram showing the successive numerical dominance of progressively more complex plants through the geologic eras and the progressive increase in the complexity of the more familiar floras.

The space assigned to the several eras (Archeozoic, Proterozoic, etc.) corresponds roughly with the length of time included in them. The Archeozoic is the earliest era, the Cenozoic the latest. The heavier lines show the increase or decrease in the number of the different kinds of plants through geologic time. The flowering plants are now (in late Cenozoic time) by far the most numerous, and their rise to dominance has been rapid; yet representatives of the older forms still persist, though they are less abundant than the flowering plants.

erous period, the cycads of the Mesozoic era, or the flowering plants of the Cenozoic era. The members of these groups became adapted to a great variety of environment and tended to occupy all the available places on the land, and some of them, such as the water ferns or the higher aquatic plants, became readapted to an aquatic existence. One group after another thus became dominant and then waned or became entirely extinct. The accompanying diagram (Fig. 1) illustrates the successive numerical dominance of different plant types and the increasing complexity of the vegetable kingdom as a whole.

Another principle is illustrated by the progressive loss of plasticity in organisms or organs as they became more complex and more highly specialized. The simpler organisms outlasted the complex or gave origin to new types, for the more complex lost adaptability to new conditions and perished during changes of environment. Most of the earlier forms of the successive groups of plants were synthetic or generalized in structure. The earliest ferns, for example, show combinations of features that subsequently became the property of different fern families, and the seed ferns combined the features of ferns and cycads.

The first simple plants, which grew in the water, probably lacked the substance commonly called leaf green (chlorophyll); they obtained their nitrogen from ammonia compounds and gained their energy by oxidation, in much the same way that some modern bacteria oxidize iron and sulphur. With the development of chlorophyll they were able to utilize directly the carbon dioxide of the air and build up complex organic compounds. The acquisition of this power of using inorganic material for food and of converting sunshine into energy marks the first progressive step in the history of plants. The second step was the occupation of the land. During the long history of land floras, covering millions of years, the two principal advances were the development of what is called secondary wood, such as forms the seasonal layers of the oaks or the pines, which enables them to increase in size for many years and carry aloft an ever larger canopy of leaves, and the development of seeds, which are a much more efficient means of reproduction than the simple single-celled spores of the lower plants.

Geologists construct their history in much the same way as any other historians, but instead of dealing with written documents or the handiwork of man they deal with the series of rocks that make up the crust of the earth, and especially with the fossils preserved in the rocks—the remains of the plants and animals that were alive when the rocks were being deposited as mud or sand. (See geological time table on page 160.)

You might suppose that this record of the past would be so scanty and broken that we could not read it. It is, indeed, far from complete, but when we remember that even the formation of a single bed of sandstone or of clay consumed a long time we can see that innumerable plants and animals might have been covered up by accumulating sediments and so well preserved that we could use them in our study of the earth's history.

The earliest chapter of the world's organic history we call the time of ancient life, or the Palaeozoic era, which is saying the same thing in Greek. This Palaeozoic era we divide into periods, each marked by distinctive types of fossils. The names of the periods that make up the Palaeozoic era, given in order from the oldest to the youngest, are Cambrian, Ordovician, Silurian, Devonian, Carboniferous, and Permian. Some of these names are geographical, each derived from the name of some place where rocks of that age are exposed. Cambrian is from Cambria, the Latin name of Wales; Silurian is from the name of a tribe—the Silures—which in Roman times inhabited that part of Britain where the Silurian rocks

GEOLOGICAL TIME TABLE

(The older areas, periods, etc., are at the bottom; the later are at the top)

Cenozoic era Modern life: mammals and flowering plants (3 to 5 million years)	Quarternary period	Pleistocene epoch	Time of the Ice Age and of the ancestors of man. Extinction of many large animals and trees. Evolution of herbs. Elevation and extension of continents.
	Tertiary period	Pliocene epoch	Cosmopolitan forests.
		Miocene epoch	Zenith of development of forests.
		Oligocene epoch	Culmination of Eocene types.
		Eocene epoch	Modernization of flowering plants.
Mesozoic era Middle life: reptiles, cycads, and conifers (5 to 10 million years)	Cretaceous period		Earliest palms. Beginnings of forests of the ancestors of the flowering plants mixed with survivors of the older Mesozoic ferns, cycads, and conifers.
	Jurassic period		Widespread warm seas, marine mammals and terrestrial cycads and conifers. Toothed reptile-birds.
	Triassic period		Land extension and shallow seas and lagoons. Red deposits. First mammals.
Palaeozoic era Early life: fishes and flowerless plants (20 to 25 million years)	Carboniferous period	Permian epoch	Dwindling of ancient forms; rise of cycads.
			Swamps of the coal age. Ferns and seed ferns, giant club mosses, and horsetail rushes. Rise of primitive reptiles.
	Devonian period		First abundant fossil land plants. First amphibians.
	Silurian period		Rise of land plants, lung fishes, and scorpions.
	Ordovician period		Rise of shelled animals.
	Cambrian period		First abundant fossils. Marine plants. Dominance of trilobites.
Proterozoic era (25 million years)	Cellular plants and primitive, mostly soft bodied marine animals.		
Archaeozoic era (50 million years)	The first life.		

may be seen; Permian is from the province of Perm, in Russia. Or the name may have been suggested by the character of the rock, as Carboniferous, a term applied to the rocks of the coal age.

After this era of ancient life, the Palaeozoic, came the Mesozoic era, a time in which the forms were intermediate between the old and the new. The Mesozoic era is divided into three periods—the Triassic, the Jurassic, and the Cretaceous. The Triassic is so named because in that period three principal kinds of rock formations were deposited in southern Germany; the Jurassic is so named because the rocks of that period are very conspicuous in the Jura Mountains; and the Cretaceous gets its name from the fact that its characteristic rock is chalk (*creta* in Latin).

The Mesozoic era was followed by the Cenozoic, the time of modern life. In the rocks of this era we find the remains of warm-blooded animals and flowering plants, and in those of the later part of the era we find the skeletons and flint implements of ancient man.

Many fossil seaweeds are scattered through the older rocks, but the first land plants found in abundance as fossils lived in middle Palaeozoic time (Devonian). (See table on page 160.) Some of these may be considered transitional between seaweeds and true land plants (Fig. 2). Others were synthetic forms combining features of organization which during subsequent ages became segregated and characteristic of separate orders of plants. Such an ancestral plant is Hyenia (Fig. 4) which combines features of the later club mosses and horsetails. Others by their complexity indicate a long period of terrestrial existence. Some of these Devonian plants are true seed ferns (Figs. 3 and 4); others are arborescent club mosses, which combine the features of plants of later Palaeozoic time (Fig. 5). Another later Devonian type, widespread geographically, is a plant (*Archaeopteris*) whose classific place is uncertain, for we cannot yet be sure whether it is a true fern or a seed fern.

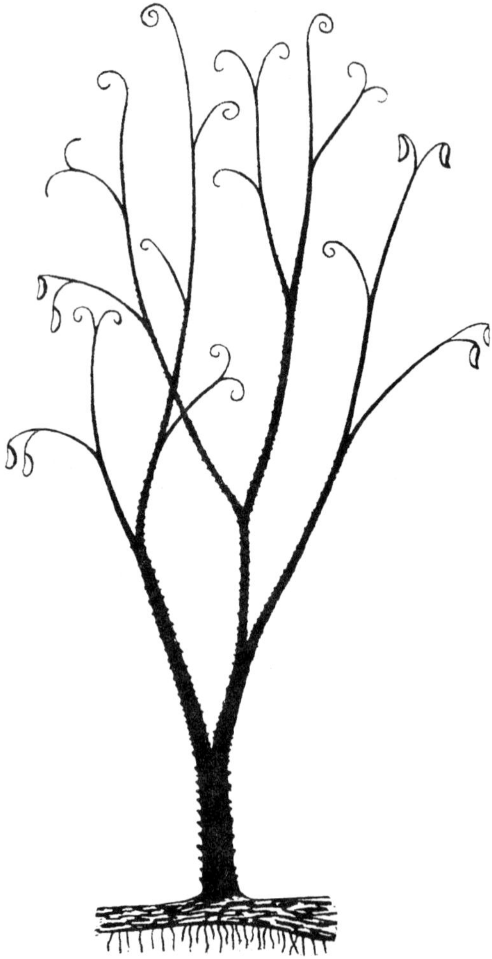

FIG. 2.—Restoration of *Psilophyton,* an early Devonian terrestrial plant showing many features suggestive of algal ancestry.

During the remainder of Palaeozoic time the terrestrial vegetation consisted essentially of the coal plants—club mosses as large as trees, with woody stems and complex reproductive structure, tree-like horsetails, a great variety of seed ferns, some ancestral to the later cycads, and trees like the modern ginkgo. The most abundant of the latter were tall trees, somewhat like modern conifers but with a larger pith in the columnar trunk and large leaves like those of a corn plant. These had

FIG. 3.—Restoration of *Eospermatopteris*, the earliest known fern, from the middle Devonian. (After Goldring.)

curious flowers and seeds quite unlike those of modern plants. Their many varieties are collectively known as *Cordaites,* and a restoration of one is shown in Fig. 6. Near the end of the Palaeozoic era we find the coal plants dwindling in number, as a consequence of the changing conditions of Permian time, and new types making their appearance, such as cycads and coniferous trees, ancestral to Mesozoic forms. The late Palaeozoic rocks of Australia, India, South Africa, and South America give evidence of widespread glacial ice. The rigors of this time in these regions expelled many of the members of the earlier cosmopolitan flora and introduced a number of new types, known collectively as the Glossopteris flora. (Fig. 7.)

The earlier part of the Mesozoic era was a time of widespread seas; the land deposits then laid down contain few fossil plants. The oldest Mesozoic rocks containing a representative flora are those laid down near the end of the Triassic period. In the long time that had elapsed since the Permian epoch many changes had taken place. A few surviving stragglers of the old order lingered on, but many of these older Mesozoic plants were the diversified descendants of the conifers, cycads, and ginkgos, though they included numerous ancestral representatives of most of the modern families of ferns. The Mesozoic has been

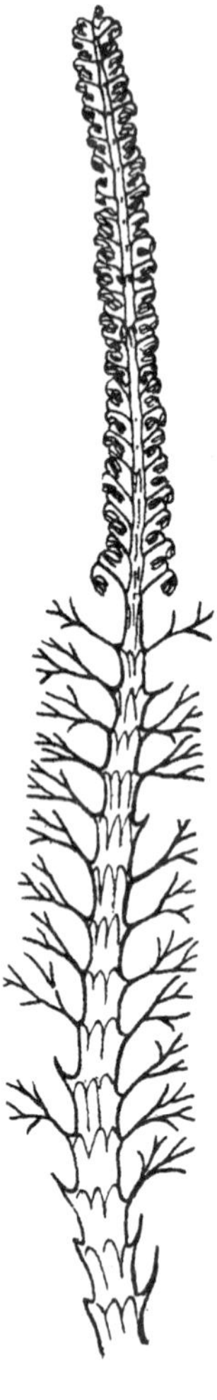

FIG. 4. — Restoration of *Hyenia,* a middle Devonian plant, which combines features of the later club-moss and horsetail lines of evolution and suggests certain features of the fern line. (After Kräusel and Weyland.)

called the age of gymnosperms (plants with naked seeds, such as the pines), but it may perhaps be more properly called the age of cycads (Figs. 8 and 9), for its rocks contain cycad-like plants in great abundance and variety.

The known Jurassic floras, whether of swamp or upland, consisted primarily of ferns, cycads, and conifers. The ferns were all forms of moderate size. None of the cycad-like forms that are so characteristic of that age of the earth's history were tall; probably none were as tall as an old cycad of to-day. Rising above the general low level of these cycads were the various conifers,

FIG. 5.—Restoration of *Protolepidodendron,* an upper Devonian ancestor of the subsequently differentiated forms of *Lepidodendron* and *Sigillaria,* the arborescent Palaeozoic club mosses. (After Berry.)

a, Lepidodendron-like leaf sears; *b,* Sigillaria-like leaf sears; *c,* leaf, about natural size.

among which were the Jurassic forms of the maidenhair tree (Ginkgo) which is to-day represented by only a single species.

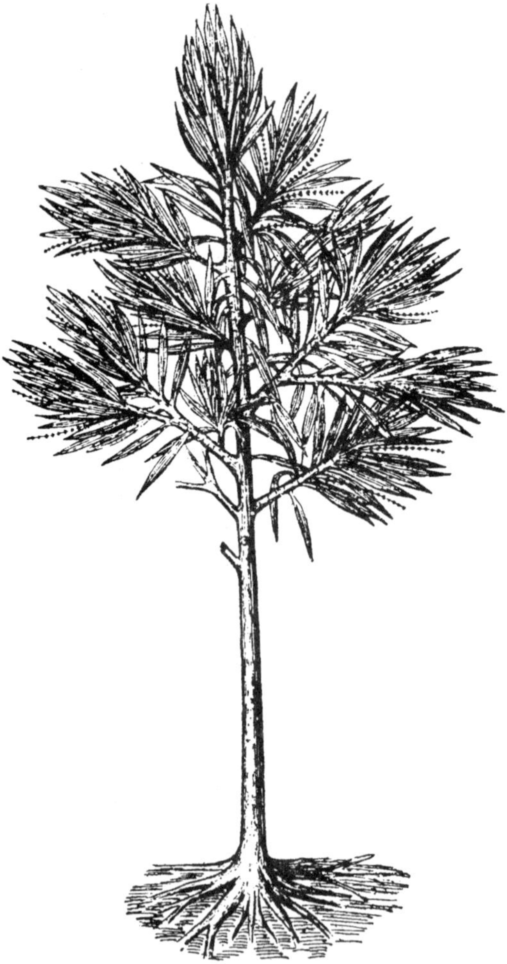

FIG. 6.—Restoration of *Cordaites*, a primitive conifer of the Carboniferous period. (After Scott.)

Lower Cretaceous plants are found in the rocks of all the continents, and they are particularly abundant in North America and Europe. The two most extensive Lower Cretaceous floras are those preserved in the Potomac group of rocks of Maryland and Virginia and those of the rocks of the opposite side of the Atlantic, in southern Portugal. Comparisons of these floras shed light on the place of origin and the migrations of the various types. A third large

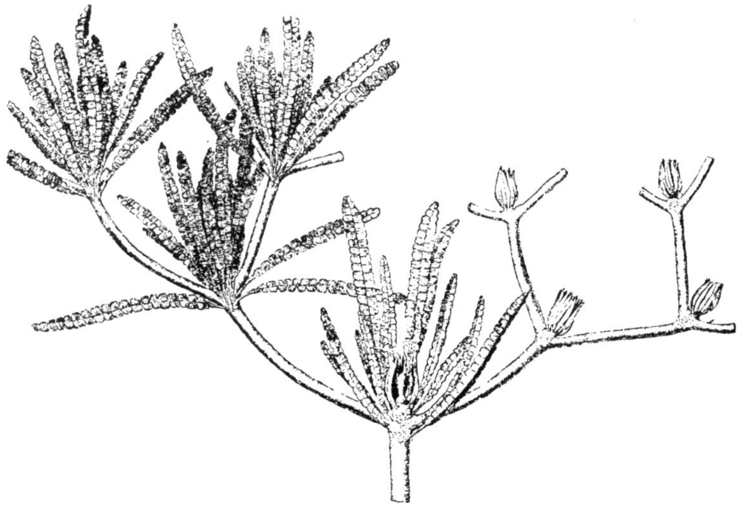

FIG. 8.—Restoration of *Wielandiella,* one of the best known branched cycads of the older Mesozoic. (After Nathorst.)

Lower Cretaceous flora is that of the so-called Wealden of England, Belgium, and Germany. Other floras of this age are found in South Africa and eastern Asia, as well as in Spitzbergen, Australia, New Zealand, and Greenland.

Although the known floras of the Lower Cretaceous epoch necessarily represent only a small percentage of the species that clothed the earth during that time, they furnish some suggestive data concerning the march of vegetation during the time in which the flowering plants first appeared, when the transformation was made from a Jurassic to an Upper Cretaceous and essentially modern flora. In the varying

FIG. 9.—Recent Japanese cycads, showing the character of the vegetation in Jurassic time. (After Wieland.)

proportions of its main types of plants, the Lower Cretaceous flora discloses local differences of soil, altitude, humidity, and precipitation. The dominant late Jurassic types—the ferns, cycads, and conifers—continued without marked change through early Cretaceous time. The early Cretaceous cycads were essentially the familiar types of later Jurassic time. They were abundant in genera, species, and individuals, and they were quite as dominant an element of the lower Cretaceous floras as they had been of those of late Triassic and Jurassic time. Before the end of the Lower Cretaceous epoch, however, most of these plants had become extinct. In rocks laid down near the end of that epoch we find preserved the first representatives of the flowering plants (the angiosperms—that is, plants having enclosed seeds, such as the walnuts, oaks, and maples), and during Upper Cretaceous time these plants gradually became predominant.

Although the seas were widespread in early Mesozoic time there were many large areas of land, but we know nothing about the floras of these areas, which may have been the scene of the evolution of the flowering plants. Certainly during late Cretaceous time they spread continuously southward in Europe, North America, and Asia, and almost everywhere the same forms occur, alike in Bohemia, Alabama, or Sakhalin Island, localities suggesting their northern origin. During Upper Cretaceous time they penetrated far into South America, reaching Argentina, and they even reached Antartica (Graham Land). These Upper Cretaceous floras invariably show a mingling of temperate and tropical types, indicative of a humid warm-temperate climate, and they all contain forms that are to-day largely confined to the Southern Hemisphere. Throughout Upper Cretaceous time new types continued to appear and the stragglers from older floras gradually died out, so that by the dawn of the Tertiary period most of the archaic forms had become extinct.

The flowering plants possess for us a profound interest,

because they yield the concentrated foodstuffs that made possible the evolution during Tertiary time of the mammals—the horses, cows, hogs, sheep, etc.—on which depend our agriculture and consequently our civilization.

The earliest floras of the Cenozoic era—the age of mammals and of flowering plants—are marked by a great modernization of forms. They consisted in large part of ancestors of forms that exist today, and their chief scientific interest lies largely in the great differences in geographical distribution which they show in contrast with the present distribution of their descendants. The contrast between the continents was not so great as in earlier times, and the whole Northern Hemisphere was clothed with forests much like those that survive today in southeastern Asia and southeastern North America. Species of magnolia, sequoia, walnut, and sassafras were then native in Europe, and during early Cenozoic time the nipa palm, the date, the cinnamon, and the bread fruit tree lingered in our Gulf States.

Gradually these floras become more modern; herbaceous plants—those having no persistent woody stem—multiplied, and then came another change of climate, during the epoch known to geologists as the Pleistocene. Because of the widespread glaciation which gives this epoch a distinctive place in geological chronology, it is often called the Ice Age or the glacial epoch, although a similar period of climatic rigor, already mentioned, occurred in Permian time, and evidence of other glacial epochs in early Paleozoic and pre-Paleozoic time has been discovered. Pleistocene glaciation was contemporaneous with the evolution of the human stock and exercised a profoundly modifying influence on the noble races of mammals and forest trees of the Northern Hemisphere. It also modified greatly the topography, producing numerous lakes, ponds, and bogs. The freshness of the deposits it left—its moraines, its bowlder till, and its sand plains, all scarcely modified in the relatively few thousands of years that have elapsed since the last ice sheets disappeared—emphasize the nearness of the great glaciers to the period of human history.

At the beginning of Pleistocene glaciation the flora of all three of the continents of the Northern Hemisphere was essentially similar. The retreat of the last ice sheet left an impoverished flora in Europe and two great asylums of survivors in eastern North America and eastern Asia. The explanation of this difference is, broadly speaking, very simple. In America and Asia, with their extensive coastal plains and north-south mountain chains, there were no insuperable barriers to the dispersal of plants southward, away from the frozen lands, but in Europe the mountain ranges (the Pyrenees, Alps, Carpathians, Balkans, Caucasus), which trend east and west, and many of which were themselves lofty enough to be local centers of glaciation, formed impassable barriers to plant migration, and branches of the sea effectually stopped the gaps between the mountain systems. Hence many of the plants of the Pliocene forests of Europe were unable to escape extinction.

Great sheets of ice accumulated over the land during at least four separate epochs. Each of these epochs lasted 10,000 to 20,000 years, and they were separated by long epochs of genial climate, known as interglacial epochs, each lasting for thousands of years, during which the floras spread northward, even to points beyond their present range. Many such interglacial floras are represented in deposits in Europe and have been diligently investigated in connection with the economic study of peat bogs. The best known interglacial flora of North America, where the extensive peat resources have been almost neglected, is that found in the Don Valley, near Toronto, Canada. Here are found impressions of leaves and other parts of the sycamore, maple, osage orange, and other types that do not to-day quite reach that latitude. Other traces of Pleistocene floras are found in cave deposits, associated with fossil remains of animals, some of which are now extinct. Swamp deposits that were overwhelmed by sand during changes along the coasts yield many species of plants, most of which still exist, such as the bald cypress, loblolly pine, sycamore, poplar, hickory, river birch, and several species of oaks. All these fossils show that the interglacial floras scarcely differed from those of to-day except in the details of distribution of the species. During the periods of glaciation these temperate forests retired southward and gave way along the ice front in this country to arctic willows and dwarf birches, which reached southward to about latitude 40°.

The post-glacial amelioration of the climate, the opening to occupation by plants of areas that had been covered with glaciers, the mixing of soils through the action of the ice, all combined to stimulate the evolutionary activity of plants, particularly the herbaceous forms. It seems probable that the herbaceous families that are characteristic of the Temperate Zone originated at this time.

Possibly more potent than natural causes in modifying the character and distribution of the existing vegetation has been the work of man, which includes the action of fire, the ax, and domesticated grazing animals. The forests are now waning. Human intercourse results in surprising feats of plant distribution, such as are shown in our familiar cosmopolitan weeds. Insect and fungal pests are similarly spread, both rapidly and widely, and tend increasingly to restrict or even to exterminate the native vegetation.

THE STORY TOLD BY FOSSIL PLANTS

The long procession of changing forms has not yet come to a halt, and man, having learned some of Nature's methods, has so applied them as to produce marvellous new varieties of flower and fruit, new habits of growth, and new adaptability to environment.

We have seen in our brief survey of the floras of the past that they illustrate the evolutionary principles set forth. We observe a gradual transformation from simple and generalized to complex and specialized forms. We see different groups becoming specialized in various ways and attaining dominance for a time, and eventually we see those that were less perfectly adapted to survive going down in competition with those that were more perfectly adapted. At one time it may be the Palaeozoic club mosses, whose trunks were mechanically defective as compared with the trunks of the contemporary exogenous conifers. At another time we see the seed ferns, with their large and complex seeds, replaced by plants having simpler and more efficient seeds. In one way or another the story repeats itself through millions of years of history.

EDWARD WILBUR BERRY

VOICE OF CREATION

WE'LL SOON BE ON AN EVEN KEEL

I feel in my heart that mankind has come to the understanding that they themselves are beasts or rather man is on equal footing with all creation. Nothing seen or unseen has been overlooked. Soon we will communicate with creation with a silent tongue and no longer be in a state of bewilderment concerning life and the Universe. We will mentally be able to travel the gauntlet of being with full confidence and understanding; understanding our beginning from being born of the Spirit into life and then reborn through death.

It is before the birth into life that will be our concern. The world of life has produced a Creation which through self analyzation and research has produced for mankind an understanding of His place in nature: Not above or below any of the seen or unseen we exist with. Too, our constant endeavor through all religious beliefs has produced a God that we can sincerely believe: It has proven that we are not left to wander the trail of life but has given us a definite path to follow and a means of preparing ourselves to cope with the future. Sometimes when our mind and body grow weary we wonder; by taking brief periods of rest we arise to thicken the foundation on which we have founded our status. Fear has never been a benefit to life as a whole, but only to individuals who have exploited it as means of power and control over their fellows in creation. Faith is the answer for the future: Confidence has been gained through knowledge of our past.

BILL HUTH

BUTTERFLIES AND MOTHS AS EVIDENCE OF EVOLUTION

ON the wings of butterflies nature writes, as on a tablet, the story of the modifications of species. . . . As the laws of nature must be the same for all beings, the conclusions furnished by this group of insects must be applicable to the whole organic world." (H. W. Bates.)

In spite of my title and my quotation from the great naturalist of the Amazon I must at the outset consider for a moment the evidence by which the belief in any scientific theory is justified. Why do we believe that the theory of the movements of the planets and satellites in our solar system is true? We believe it because by the light of this theory astronomers can predict the future, and we know from experience that their predictions will be verified. Farseeing people years ago made arrangements for observations on June 29, 1927, because they had been told by astronomers that on that day the moon would come between the sun and the earth, that its shadow would sweep across England from Southport to Hartlepool, and that every place in succession on that line would be, for about 25 seconds, buried in the darkness of total eclipse. And at the precise moment we saw, just as predicted, its sudden onset and swift passing away, while, in favored places where fortunately the astronomers had erected their instruments, the clouds cleared and unveiled the glorious spectacle of the Corona, drowned at all times except during total solar eclipse in the overpowering light of the sun.

We cannot expect to find such clear-cut evidence of evolution in the story of life on the earth, but we have abundant opportunities of applying a test that is in principle the same as that which justifies our belief in the astronomical theory.

If the theory of the movements of the solar system is correct an eclipse will take place at such a time and place; if the theory of evolution is true then certain organs or parts should be found in the ancestors of animals that do not now possess them or perhaps still exist in the early stages of animals of to-day. We search in the rocks, we find and dissect the early stages, and there the missing parts are revealed. The prediction founded on the theory of evolution is verified.

In order to convince others it is necessary to be convinced oneself, and nothing is so convincing as personal experience. Therefore I will tell of a prediction founded on evolution and its verification nearly forty years ago.

Few people, I suppose, realize that their teeth are among the most ancient of the parts that make up the human body. They are, for example, older than the hair, which was derived from the horny scales of ancestors much nearer to our own times than those which gave us our teeth. Teeth have come by direct descent from a remote ancestor that was covered with scales like a shark—scales which in this ancestor, as in the shark, passed over the lips into the mouth and, without any essential modification of structure, were used as teeth. The scales and the teeth of sharks are composed, like our own teeth, of hard dentine, developed from cells below the epidermis, or surface skin, covered with the much harder enamel, developed, at any rate mainly, from the epidermis itself. Scales of this kind and teeth have the same structure, the same development, and in the shark the one passes over the lips into the other.

Gradually, through long ages, body-scales like those of the shark were replaced by scales of different structure, and these directly or indirectly gave rise to reptilian scales, birds' feathers, and mammalian hair. But all through these changes the ancient scales in the mouth—the teeth—have remained essentially the same. Existing birds have lost them, but they were possessed by their ancestors, as we know from the fossil birds of both the Old World and the New.

The story of the mammals—our own branch—is more completely represented by animals alive to-day because of the preservation in Australia, cut off by sea from the stress and rush of life in other great land areas, of the duck-billed platypus (*Ornithorhynchus*) and the echidna, which is found also in New Guinea. Thus preserved from extinction, these remarkable animals have come down to us, descended from a link that connects the mammals with some primitive reptilian or pre-reptilian ancestor. Their temperature is much lower than that of mammals; their skeleton shows strong reptilian affinity; above all, they lay eggs like reptiles and birds, but they suckle their young by a primitive form of mammary gland. Yet these ancient forms, descendants of an ancestor through which the mammals received their teeth, are now, both of them, toothless. The echidna, feeding by means of its tongue, like the true ant-eaters of South America, is entirely toothless; the platypus has hard, tooth-like plates for crushing the insects and mollusks of the streams in which it lives. These plates are really hardened gums; they have nothing of the structure of teeth. So the evidence looked for was wanting just where we should chiefly expect to find it. The evolutionist was nevertheless confident that these animals or their immediate ancestors originally possessed teeth but had lost them, and he believed that this inference from the known to the unknown might at some time be verified.

In 1888 I was working on the hair of the platypus and was able to show that it retained scale-like features which have been lost in the higher mammals. It was desirable to examine a young specimen, and knowing that Professor Kitchin Parker possessed one I asked if he would lend me some sections of the head prepared by his son, Professor Newton Parker. Just as I was about to examine them the thought flashed through my mind, "Perhaps at this young stage the platypus has not lost its true teeth." I looked, and there they were, complete, with dentine and enamel, lying beneath the gum. The prediction was verified.

I cannot refrain from saying a few words about the generous treatment I received from that great man. I had borrowed his sections not to look for teeth but for hair, and he might well have said that I had anticipated the study he intended to make and must not publish the discovery. Far from it, he wrote full of enthusiasm and kindness, offering himself to communicate my paper to the Royal Society. It was a splendid thing for a young man to meet with so much kindness from one more than twice his age. I shall never forget it, and I hope the memory of it has enabled me to help on my younger comrades.

Later on Professor Charles Stewart found that the teeth cut the gum and are used for a time by the young platypus, but that they soon fall out and are replaced by the horny plates which invade their sockets.

Before considering the evidences of evolution furnished by butterflies and moths, I will attempt to answer the objection that nobody has ever seen one species turn into another and that nobody has brought convincing proof that species do change in this way. Just such an objection might be raised by one who paid a short visit to this planet and was assured that children became men and women. "I have been here for a whole week," the visitor might well say, "and I have looked everywhere for this transformation, but I have never seen a child turn into a man or woman." But a week is a far greater part of the period of human growth than is the time of human observation in the life of a species. Furthermore, if the visitor prolonged his stay indefinitely he would still never see a child "turn into" a man or woman, for between the two intervenes a growth so gradual that no difference is perceptible from day to day or from week to week. So is it with evolution. One species does not "turn into" another: it becomes another species through a series of gradual changes, and at no time would it be possible to say—"*Now* the change has come; what was species A yesterday is species B to-day."

To prove that species A, known to us only from remains in the rocks, had become species B of to-day it would be necessary to restore to life the animals of innumerable past generations of beings and to show that, whereas those of adjacent strata could interbreed, their ancestors (species A) could not interbreed and produce fertile offspring with their living successors (species B). As this is manifestly impossible, we infer from the gradual changes of form or structure preserved in the rocks that A is a different species from B, which has apparently sprung from A by direct descent.

If, however, we cannot witness the transformation of one species into another any more than we can witness the sudden transformation of a child into a man or woman, we *are* able to witness the results of a series of changes in living forms in adaptation to the conditions of life—

to what is called environment—and it is here that the butterflies and moths provide excellent illustration of evolution.

When I was a boy the common peppered moth was known to produce a rare black variety. The growth of the manufacturing districts of Lancashire and Yorkshire has greatly increased the volume of the smoke there, which, carried by the prevalent southwest winds, has done deadly work over a wide area, killing the gray lichens and leaving the tree-trunks dark and sooty. Resting on bark like this the peppered moth would be more conspicuous to the eye of a bird seeking food than the black variety, and accordingly for many years this black form has entirely replaced the other form in these northern tracts. The others, being more easily seen, have been eaten. And the peppered moth is not the only species that shows change; several other bark-haunting moths have also become much darker in the same strip of country and during the same short period. Furthermore, similar changes have been observed in the moths of other smoke-producing areas in this country and on the Continent. Harrison has recently shown that some of these moths have become dark after their caterpillars have been fed for many generations on plants contaminated with salts of manganese, such as are contained in smoke. Inasmuch as the effects were transmitted in Mendelian proportions we must conclude that the salts acted upon the germ cells.

A still better but less well-known example of change in colour is found among the butterflies of tropical America. In each district these insects and some of the day-flying moths form groups that are of similar pattern and colouring but that are composed of species having very different degrees of relationship. Among the groups of any locality one species is generally predominant in numbers and is among the most conspicuous. But as the naturalist passes from one district to another he sees that the pattern of the groups becomes different, "as if at the touch of an enchanter's wand," to use the words by which Bates describes the change that sweeps over all the diverse members of a group. In Central America and Venezuela the chief group is made conspicuous by a tawny and black-barred pattern; in the Guianas the members of the corresponding group have much darker hind wings—the wings of some are almost entirely black; across the Amazon, in eastern Brazil, the hind wings change in the opposite direction, gaining a bright yellow stripe; high up the Amazon, at Ega (now Teffé), Bates' headquarters for many years, the general colour of both wings changes from tawny to deep chestnut.

How shall we explain the advantages of these local colours and patterns, which run through many distantly related species? Let us see.

The insect-eating animals of each district, especially the birds, learn by experience that insects having certain conspicuous colours and patterns have an unpleasant taste or smell or are indigestible. So if a number of different noxious species bear the same pattern the birds easily learn to avoid them, with little waste of insect life in experimental tasting. Those so marked survive. Hence the great advantage to the butterflies of a combined advertisement or announcement that they are unpalatable, instead of each distasteful species having its own warning pattern, requiring to be tested separately. It is probable, too, that among the members of a large group there are many degrees of distastefulness and some also that are not distasteful at all, but that flaunt a false advertisement and live on the reputation (or rather the disreputation) of the others. This false advertisement was interpreted by H. W. Bates and is spoken of as

Batesian mimicry. The combined advertisement was interpreted some years later by Fritz Müller and is called Müllerian mimicry.

I hope that the above paragraph makes clear the advantage that would be gained by a species whose pattern resembled that of another species which was not eaten by birds. Probably we should have an excellent opportunity of actually witnessing evolutionary progress if we could compare one imperfect mimic in such a group with what it was a hundred years ago. A century is only a drop in the ocean of time, but even in this short period some notable change might be evident. Well, we are fortunate enough to be able to make this comparison.

Between the years 1825 and 1830 the great naturalist William John Burchell was travelling in eastern Brazil, making extensive collections of animals and plants, concerning which he made the most accurate and detailed notes. I only wish that all naturalists to-day would do their work as well. His whole collection of insects is in the Oxford University Museum, and among them are many butterflies belonging to the great yellow-banded group of which I have spoken. In a certain species known as *Lycorea halia,* the yellow band along the hind wing is not so bright as in other species of the group, and the eight specimens in the Burchell collection show that this characteristic feature was on the average even less bright a hundred years ago. It may be suggested that the colours have become darker with age; but with age butterfly pigments generally become paler rather than darker. Besides, among the eight there is one that is fairly bright, though by no means equal to the brightest of to-day, whereas among specimens recently caught a small proportion resemble the seven that are not so bright. The evidence that the colour has changed slightly in a hundred years is, I think, unquestionable.

The naturalist does not require such evidence of evolution as that described above. If future enquiries should prove that it and similar evidence are based on insufficient material and are hence illusory his faith in evolution would remain unshaken. Why is this? Because during the whole of his life's work he is always meeting with facts which, on the theory of evolution, receive a clear and fascinating interpretation, but which without it are meaningless. Accept evolution and they fit into their place in the scheme of things; reject it and they are isolated and devoid of interest. An instance or two will make this clear.

The female of the common vapourer moth has lost the power of flight. The brown male, which has a white spot on each forewing, flies actively by day and may be seen dashing through the streets and squares of London seeking the female where she sits quietly on the outside of the cocoon from which she emerged. Later on she will lay her eggs on this cocoon and die without leaving it. The little caterpillars that are hatched out eat many kinds of plants and are in no danger of starvation, for a short journey will bring them to food. Therefore the female does not require wings in order to seek the plant that provides food for her offspring and lay eggs on it, and she does not have to seek her mate. All the seeking is done by him. He will even enter a house and creep under a door to enter a room where a freshly emerged female is being kept in a box. Now the evolutionist knows that this nearly wingless female is descended from ancestors that possessed wings like other moths and that her rudimentary wings have become what they are by gradual degeneration. Why are the rudiments there? No theory except evolution can give a reasonable answer. To believe that by an arbitrary act of creation one moth was given useful and another useless wings is a childish creed—an unworthy conception of a Creator. And such an assumption becomes even grotesque when the facts are examined a little more closely. The wings of ordinary moths are developed within the much smaller wing envelopes of the chrysalis and expand to their full size only after emergence. Before this they can fit into the narrow space only by a complicated system of pleating. But the rudimentary wing of the female vapourer lies within a chrysalis wing far larger than itself. Thus A, Fig 1, shows the outline of the chrysalis wing drawn 7 times larger and broader than its natural size, and B, similarly magnified, shows the outline of the wing of the female moth, which lies within it and remains of the same size when the moth has emerged.

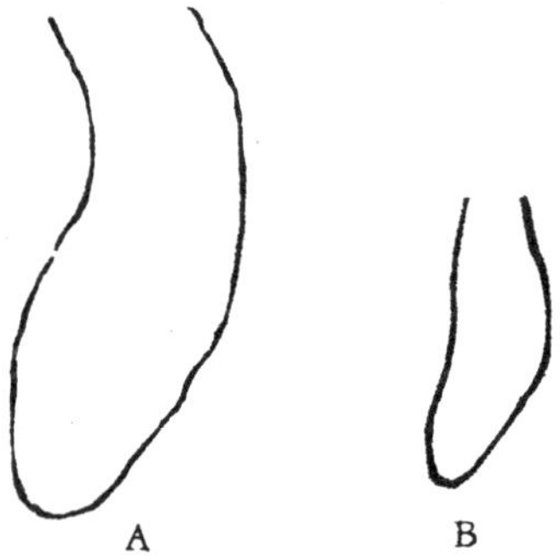

FIG. 1.—Outlines of rudimentary wings of vapourer moth.

To the evolutionist these facts mean that the useless wings, being probably a source of danger to the moth as well as a waste of material, have been gradually reduced by natural selection until they became first no longer and finally much smaller than the chrysalis wing cases. These chrysalis wing cases are also themselves reduced but, concealed in the cocoon and less subject to selection, their shrinkage is not nearly so great.

The most wonderful instance of butterfly mimicry that we now know is the "swallowtail"—*Papilio dardanus*—of Africa and the neighboring islands. (See Fig. 2.) The male is a pale-yellow black-marked butterfly, having, like most swallowtails, long tails to the hind wing. A male from Madagascar and one from Uganda are shown in the accompanying illustration (Fig. 3). In Madagascar and the Comoro Islands the pattern of the females is very like that of the males, and they also have tails. On the mainland of Africa, however, male-like females are known only in Abyssinia. In other parts of the continent the males are pale-yellow black-marked butterflies with long tails, but the females are entirely different, resembling quite different tailless butterflies that have an unpleasant taste and that bear conspicuous "warning" patterns. The commonest of these mimicking females is a black-and-white tailless form (4), and the butterfly which it resembles and is therefore called its "model" is represented in 5.

Now the evolutionist felt confident that these tailless mimicking females were derived from females that had tails like the males, and his confidence has received a three-fold verification.

The first verification was obtained about twelve years ago, when Dr. W. A. Lamborn discovered that the female chrysalises have pockets for the tails, although no tails are developed within them.

The second verification is found in the fact that underfeeding the caterpillar or subjecting the chrysalis to cold may result in the production of rudimentary wing-tails.

The last and most convincing verification is provided by the Abyssinian race of the swallowtail, in which the females are generally male-like, but some comparatively rare females have gained the mimetic pattern yet have not lost their tails. An example is shown in 6. The right tail is well developed, although the left one has been torn off, perhaps as a result of attack by some enemy. The specimen figured is one of five—two in the Prague Museum, one in Lord Rothschild's Museum at Tring, and two at Oxford, the second having unfortunately lost both its tails. The figured specimen was captured by Mr. Arnold Hodson in southwestern Abyssinia on November 15, 1925. Four of the models, of which one is represented in 5, were taken in the same locality on the same day.

It should be added that other nauseous butterflies are mimicked by other female forms of the same swallowtail, as well as by very different butterflies and by day-flying moths. Four out of the five Abyssinian mimetic females resemble 6, but the fifth, at Prague, exhibits the very different colouring of another model.

Examples could be multiplied indefinitely, but I believe that those here described afford sufficient evidence that predictions based on evolution are verifiable and have been verified, and that natural history becomes in the light of evolution a living and inspiring study.

Although there are widely different opinions about the causes of evolution, it is probable that no living student of nature has any doubt about the *truth* of Evolution.

EDWARD BAGNALL POULTON

MAN WAS THERE AT THE BEGINNING

We must understand that our teachers, religious leaders, fellow man and scientific generations have continually presented humankind with made over words and names which they have used to explain to us how we came to be. We have been developed through the process of evolution not as a product achieved through evolution. We have been since the beginning. As a matter of fact I feel in my being that we need to worship Creation instead of all of the Holy Deities we have been offered. After all, what Creation is, is nothing to fear.

All of this great life-being is and has always been one. Evolution was not a cause to create man; man has been since the beginning. Through the different stages of design humankind, including all of creation, have manifested themselves into what we should realistically understand Him to be: A GOD TO HIMSELF.

Nothing can exist alone. All Creation is the stew! We're endeavoring steadily towards a goal of greener pastures - a haven or heaven of peace and tranquility. This will never be achieved or understood until we unite and grasp the fact we equally share with all the compassion of life, at peace.

A VOICE OF CREATION

THE EVOLUTION OF THE BEE AND THE BEEHIVE

For so work the honey-bees,
Creatures that, by a rule in nature, teach
The act of order to a peopled kingdom.
They have a king and officers of sorts,
Where some, like magistrates, correct at home,
Others, like merchants, venture trade abroad;
Others, like soldiers armed in their stings,
Make boot upon the summer's velvet buds,
Which pillage they with merry march bring home.

Shakespeare.

In a primitive and savage state of society each individual of a tribe is a host in himself. He is at once a

Tinker, tailor, soldier, sailor,
'Pothecary, ploughboy, thief,

and except that he cannot very well be his own undertaker he performs all the functions of the various traders and professional experts that in a more civilized state of society are carried on by numerous men, each suitably trained for one pursuit, and generally for only one. The North American Indians built their own wigwams, tilled the soil, fished, hunted, fought in tribal wars, and engaged in other activities. The women took a large part in the drudgery of life—cooking, tending the young, helping in shifting the camp. But as affairs became more complicated a higher social order was established. The medicine man, who acted both as priest and doctor, was evolved. A chieftain was set up. The old men became counsellors. Still each family lived in its own wigwam and not with others in an apartment house or a hotel.

Certain social communities other than human have in a similar way evolved from simple beginnings, and one of the highest of these is undoubtedly that of the honey-bee; and the society of the honey-bee is even more complex than anything in our own civilization. "The bee in its own line," writes J. A. Thomson, "is hardly inferior to man, and represents an achievement that angels might desire to look into."

FIG. 1.—Queen bee.

FIG. 2.—Worker.

FIG. 3.—Drone.

In a beehive there are three ranks of individuals. First, there is the queen bee (Fig. 1), who is indeed the mother of her people, for she alone lays eggs; and as a rule she is a solitary monarch, and tolerates no rivals. Then there are the workers (Fig. 2), which in structure are females, though they have ceased laying eggs. Like Martha, they are cumbered with much serving. Third, there are the males, or drones (Fig. 3), quite useless in the conduct of the affairs of the hive except that one of them will ultimately fertilize the queen bee.

The beehive itself is a very complex affair. When a cluster of bees have swarmed (Fig. 4) they take refuge in some cavernous structure, such as a hollow tree; or maybe they are enticed into a skip, or hive, by a beemaster. Then their first task is to clean out their new home, smoothing out the walls, and next the worker bees begin to produce wax. A row of them hang on to the top of the hive, and they support a second row, and these a third, and so on until they have a living string or network of bees hanging from the roof. All these bees are producing wax, and in order to do this they must be fed on honey. The wax is secreted by the hinder part of the bee, pressed forward by the legs, and shaped and moulded by the jaws (Fig. 5). Parallel with this veil of bees will be a second and a third, and maybe more; and every member of the veil is passing wax forward up to the top, where a waxen foundation for the honeycomb is being formed. As soon as a stout foundation has been laid the veil of wax workers breaks up and the bees begin to work independently of one another. Now they add their film of wax indiscriminately to one or another part of the comb.

FIG. 4.—Bees swarming.

The whole of this procedure seems thoroughly unorganized. None of the bees have ever seen a honeycomb before. They are all working in complete darkness. They have no one to direct them, no foreman or master builder, yet so accurate are the results of their work that the cells they make are of uniform size and are so arranged that each cell is hexagonal in cross section—and a six-sided structure contains more space with the use of less material than a structure of any other shape. There are about 9,000 cells in a square foot of honeycomb. The cells are nearly all of the same size and serve as the homes of the workers. Somewhat larger cells house the drones, and other deeper cells are used for storing pollen or honey. So accurately is each comb placed with regard to its neighbour that the space between them allows only two working bees to pass each other (Fig. 6) as they carry on their ceaseless labor.

FIG. 5.—Under surface of a worker bee, showing the hind legs pushing out from a pocket a flake of wax, which will be passed forward to the mouth and kneaded into the cell of the comb. (After Casteel.)

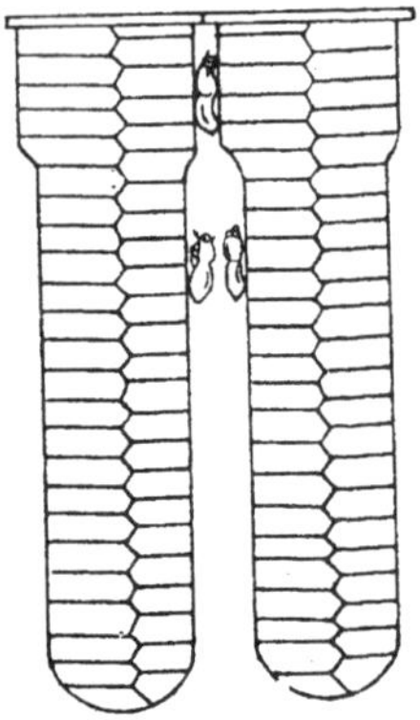

FIG. 6.—Two neighbouring combs of a honey-bee's hive, showing the shape of the ends of the cells and the space between adjacent combs, which is just wide enough to allow two bees to pass. At the top, where there is no need for bees to pass, are the larger cells in which honey is stored.

The cells of the drones, having to accommodate a rather larger larva, are made slightly bigger, and in some wild honey-bees they are all placed together in a special drone comb. The cell in which the queen is reared is, however, altogether different. It is about the size of an acorn, and its walls are much thicker than those of the other cells and are usually rounded. As soon as the queen bee is hatched out these walls are destroyed and their wax is used to add more worker cells to the comb. An average hive (Fig. 7) will contain some 30,000 working bees, some 2,000 drones, and but one queen, who alone is a functional female and produces a continuous flow of eggs.

The queen moves along on her egg-laying journey, exploring every empty waxen cell with her feelers and inserting a single egg into each cell. She never seems to tire, and she

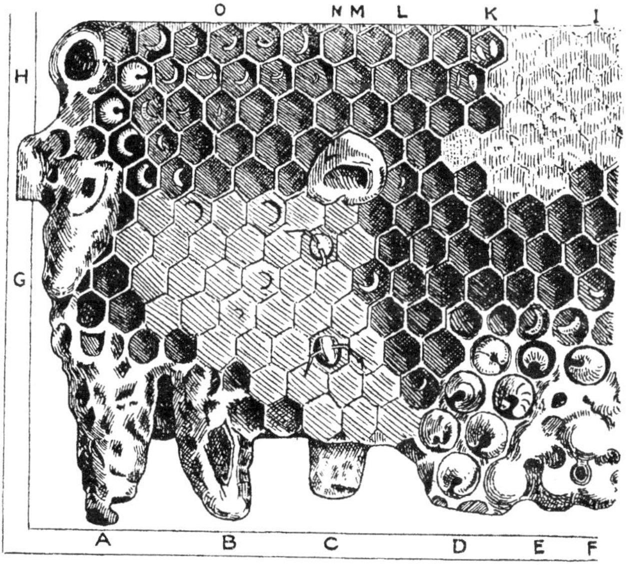

FIG. 7.—Comb of hive bee (Natural Size).

A, empty queen cell; B, the same, torn open; C, the same, cut down; D, drone larva; E, F, sealed drone cells; G, sealed worker cells; H, old queen cell; I, sealed honey; K, masses of pollen; L, pollen cells; M, abortive queen cell; N, emerging bee; O, eggs and larva. (After Cheshire.)

never misses a cell. During her progress she is surrounded by a small court of worker bees, who act as courtiers, walking backward. Some of them fan her with their wings; others stroke her with their tongues; still others feed her with half-digested pap, or "royal jelly," and all are humming most agreeably and soothingly. During May and June the queen will lay three thousand to four thousand eggs every four and twenty hours, and in the course of her life of four or five years she produces hundreds of thousands of eggs. But should the number of bees in the hive decrease she will cease laying eggs, as there will not then be sufficient workers to attend the resultant larvae.

As soon as an egg is placed in a cell the worker bees get busy. They push their heads into the cell and seem to do something to the egg, though what it is is not clearly known. Within three or four days a very small white maggot-like grub (Fig. 8) emerges from the eggshell. It has no legs and is devoid of everything we associate with insects—it has no wings, no stings, no feelers, no eyes, and its intestine ends blindly.

For the first day or two the young larvae are fed from the secretion of the salivary glands of the workers. This is known as pap, or "royal jelly." The larvae not only lap this up, but float in it. On the fourth day this food is mixed with honey, and henceforward the drones are completely weaned and feed entirely on honey and pollen. The queen bee, on the other hand, lives on nothing but royal pap. After about six days the larvae cease to feed. They are then sealed up in their cells (see Fig. 8) by the worker bees and each larva makes a cocoon case, in which it forms a chrysalis or pupa.

After a few more days the young bee emerges from the cocoon and commences to gnaw her way through the waxen covering of her cell. In this she is aided by numerous workers, who hurry up from outside, and as soon as she staggers into the darkness, the heat, and the bustle of the hive, these workers arrange her hair, clean her, and offer her honey to eat. But she has undergone a kind of resurrection and is at first bewildered, trembling and feeble. However, she soon settles down. But she does not quit the hive till nearly a week after her emergence from the cocoon. Yet all the time she is kept busy helping the older workers. When she first leaves the hive she may attempt only small flights. She has to learn her way home before she sets out to collect honey from the sugar glands of plants or pollen from the pollen sacs of flowers. She may make as many as a hundred flights a day (Fig. 9), bringing back beebread, or pollen, and honey, which are stored in separate cells and used as food for the inhabitants of the hive. Through long ages the flowers and the bees have evolved together and they are now fitted to each other as hand to glove.

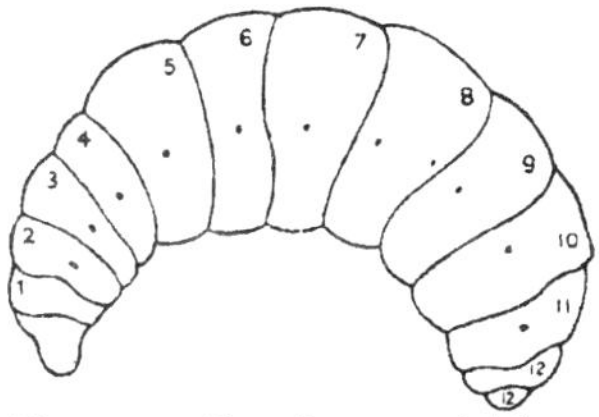

FIG. 8.—Bee larva. (After Fleischman.)

It will be observed that the life of the whole colony is based on the principles of pure socialism, and that the social system is superior to ours. There is no unemployment in a hive; there are no strikes, no lock-outs. Except the drones everyone works continuously and at high pressure. A vast majority of the bees live as workers, entirely renouncing individual rights in their effort to continue the swarm—to make sure that another queen bee may always be ready when her predecessor dies. Self-preservation and self-propagation are completely transcended that the swarm—the social unit—may be continued. Sometimes bees act as foragers, collect-

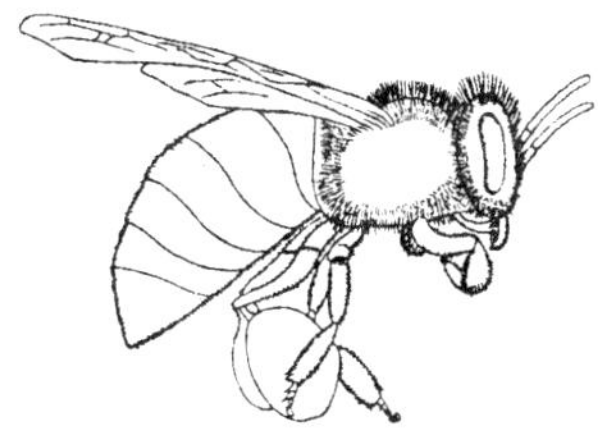

FIG. 9.—A bee upon the wing, showing the position of the middle legs when they touch and pat down masses of pollen. (After Casteel.)

ing pollen or nectar from plants to be turned into honey. Sometimes they act as chemists, as when they inject drops of formic acid into the stored food to prevent its fermentation. Sometimes they are sealing down cells. Sometimes they are sweeping and cleaning and scavenging to keep the hive clean, and dragging dead bees into the open. Sometimes they are acting as policemen to guard the hive—to scare away intruders. Sometimes they are architects and wax-workers and moulders. At times some fan their wings to ventilate the interior of the dark hive and to aid in the evaporation of the water in the honey if it is too weak. Some of them act as nurses and some as maids of honour, who do not allow the queen to get out of their sight. As has been pointed out by a learned divine:

"Three facts emerge from a study of this community:

"1. The lesson of solidarity, of the social spirit, to which the interests of the individual are subservient.

"2. The distribution of labour in accordance with the law of mutual help, each doing his work like an instrument in a vast orchestra and all producing a beautiful harmony.

"3. The law of sacrifice for the sake of the future race."

Each individual is so wrapped up in the community that if isolated from its fellows it dies. The constant sense of mutual help, of self-sacrifice for the future race, is the dominating characteristic of all bees, and there is something that Maeterlinck calls the "spirit of the hive," which in some way guides, directs, and controls the work of this strange, self-sacrificing community. Here there is no private property. As Dryden says in his translation of Vergil's book about the bee: "All is the State's; the State provides for all." In a passionate devotion to duty and in an energy expended solely for others, in a single-minded purpose, the queen and the worker honey-bee are unique among animals. Now, how has this wonderful socialistic life come about? How has it been evolved? How can we discover the steps in its evolution? We can trace the social bees and wasps back to solitary bees and wasps, and we can trace a steady growth of complexity in the habits of life of these solitary insects and in the complexity of their homes until we reach the stage that is briefly described above.

One of the characteristics of the bee, as anyone can observe,

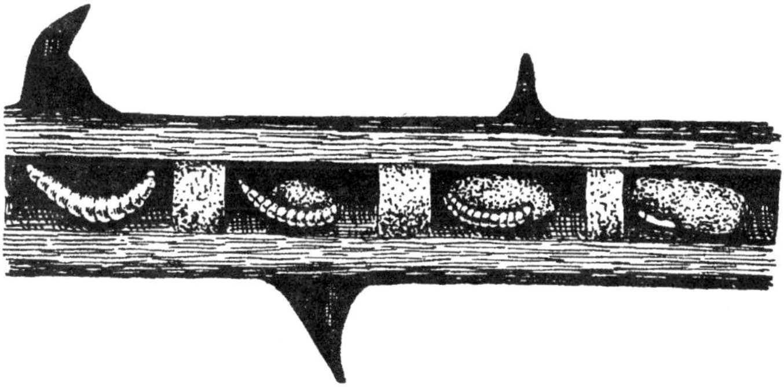

FIG. 10.—Nests of a small carpenter bee in a hollow bramble stem; showing egg, three larva in different stages, and bee-bread in three of the cells. (After Dufour and Perris.)

is a hairy body. The body is so completely covered with hair that it has a furry appearance. Now the simplest form of bee, which has no common English name but is known scientificially as *Prosopis,* has hardly any hair. Its tongue is rudimentary, its hind legs are not adapted for collecting pollen, as are those of the honey-bee, and it does not lead a social life. It makes separate cells, each lined with a silken membrane, in the stems of such plants as brambles (Fig. 10); or it burrows in the earth, or even in the mortar of walls. It collects little if any pollen and it stores in its separate cells a very weak honey, in which the egg is laid. This bee is common in America as well as in Great Britain.

Here we have a bee that has not developed the typical hairs of a honey-bee, that collects little or no pollen, that stores the cells in which eggs are laid with thin honey, which it brings straight from flowers and does not first deposit in honey cells—a bee that produces separate and distinct cells, which may or may not be in contact.

A little higher up in the scale of progress we find another

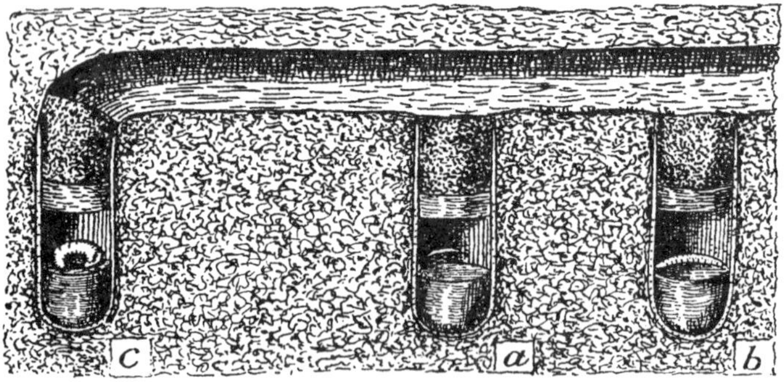

FIG. 11.—Nests of a solitary bee, tunneled in the ground. *a*, cell provisioned and supplied with an egg; *b*, cell with young larva; *c*, cell with older larva. (After Valery Mayet.)

group of bees, which burrow tunnels in sandy soil, some of them nearly a foot in length. The tunnels and the cells are lined with a paper-like material, and the cells are divided by partitions, which may or may not be in contact. These cells are furnished with a fluid mixture of pollen and honey, both of which have been swallowed by the mother bee. All this shows an advance over the work of the bee first described, inasmuch as pollen forms a conspicuous part of the food of the larvae and there is a common entrance through a tunnel to the cells (Fig. 11). The nourishing fluid is more liquid than that supplied by the higher bees, and the papery lining is formed from a slime that gradually hardens. In this group, as in many others, the male is considerably smaller than the female.

Still higher up in the scale of progress we find a solitary bee, which also burrows into the ground—in gravel paths or among grass—and also stores its cells with honey and pollen. Although these bees are in a sense solitary they live in colonies that consist of large numbers; a colony may comprise a thousand cells. The sexes differ very much in appearance and are not often found together. The bees of this group are of economic value, for they aid in the fertilization of fruit trees. The bees of one particular branch of this group construct for a number of families a common gallery, which ramifies about in the soil, and these bees thus perform a certain collective or social work (Fig. 12). But the task of constructing each cell and of providing food for the larvae is the work of one family and not the collective work of many bees.

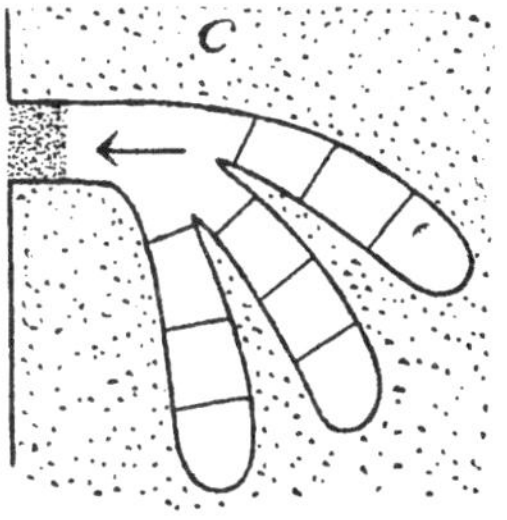

FIG. 12.—A series of end to end cells that have a common opening, indicated by the arrow. After the eggs are deposited the opening of the tunnel is closed.

Another group of bees falls under the common name of leaf-cutting bees (Fig. 13). This bee is more robust than the ordinary hive bee and has a broader head. It makes nests in hollows in stems, in wood, or in the soil. The cell is made of leaves or of parts of leaves or petals of roses and other plants, which are moulded into a thimble-like form that has a lid composed of a smaller round piece of leaf. The cells are placed end to end and not side by side, and the pieces of leaves are gummed together. The string of cells thus made rarely exceeds seven. When completed each cell is half filled with pollen, on which an egg is laid. Other species of this group enter houses in India, and both sexes there take part in making cells of clay, which may be set in any hollow tube, such as the barrel of a gun or the hollow in the back

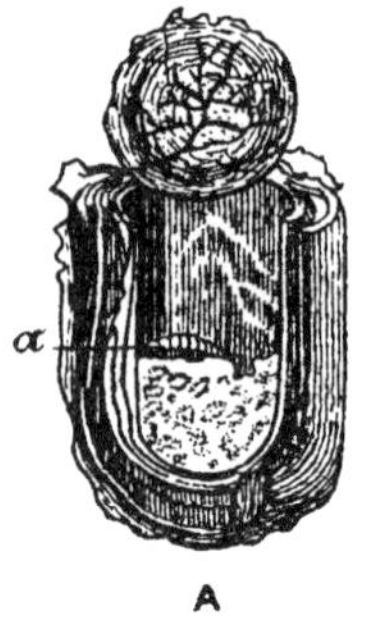

FIG. 13.—Nests of leaf-cutting bee. A, one cell separated, with lid open, and the larva (*a*) reposing on the food; B, part of a string of the cells. (After Horne.)

of a book which is lying open, or in the interior of a piece of bamboo.

Then we have the mason bees, which construct nests of sand or soil or clay moulded together with some sticky substance. Externally each cell is rough and untidy, but inside it is smooth and polished. Generally ten to twenty cells form a nest, and each cell is stored with a mixture of honey and pollen. Some of these mason bees are very hairy, and the two sexes differ from each other in colour. In its general appearance this bee is something between a humble-bee and a honey-bee, but it is solitary in its habits. Each cell may be an inch deep, and here we see pollen being carried on the hairs of the under surface of the body. To place this pollen in the cell the bee enters backward and, with the aid of its hind-legs, brushes and scrapes and combs the pollen off from the under surface of its body so that it falls into the cell. This is a distinct advance on what we had at the beginning of our series, where the pollen is swallowed and brought up again. The pollen and honey are, however, not kept separate, but are worked up by the jaws of the bee into a paste, on which the egg is laid, and the cell is then closed with cement. The work of building this cell takes about two days, and after it is finished the bee will begin to make a second cell close to the first, and will continue its work until it has made eight or nine cells, when it places a thick, dome-like layer of mortar over the whole series. The result is a nest about the size of half an orange. The larvae live in these nests for months; they do not pass through their life-history so rapidly as the honey-bee.

An equally ingenious insect is the carder bee, which has developed the habit of making nests of wool or cotton, obtained from plants that grow in the neighbourhood. This bee is referred to by Gilbert White in his "Natural History of Selborne." The male, like that of the honey-bee, is conspicuously larger than the female. These carder bees build their nests in any hollow, such as a cavity in wood or a deserted nest of other bees, or in an empty snail shell. In order to retain in the cell the fluid mixture of pollen and honey they line the cell with a thin cement. A few allied species form their cells of resin instead of wool or cotton.

The last of the solitary bees we shall consider are the carpenter-bees. These are big, burly black or bluish-black bees. They have powerful jaws, with which they carve their way into dried wood. They avoid living timber, but they will bore a hole into a beam or a rafter, and this hole will lead into three or four parallel galleries, in which they place their broad cells. Between the cells they make partitions formed of fragments of wood cemented together by their saliva. These bees pass the winter in the adult stage, hibernating in the imago condition. Both sexes reappear in the spring, and some species may take two years to complete the cycle of their life history. They are very hairy, and some of the females closely resemble the bumble-bee. The cells are provisioned with pollen, and the bees apparently produce little or no honey.

If we now turn to the social bees we find three groups. One is known as the mosquito bee, from its very small size. These bees are also sometimes spoken of as stingless bees, though they have a rudimentary sting, which they do not use. Little is known about them, but they form communities consisting of a large number of individuals. We do not certainly know whether these bees are all the product of a single queen or whether there may be more than one egg-producer in each colony, but the evidence seems to show that every colony has its own queen. The nests are rich in honey, and to prevent them from being robbed the workers, who are usually occupied in collecting pollen, also collect clay, with which they build a wall to protect the nest, which is generally placed on a bank or in the trunk of a tree. Every nest is thus completely surrounded with clay. The honey is stored in separate cells or in clusters of cells, each cluster about the size of a pigeon's egg, and these are placed at the bottom of the hive, away from the cells where the larvae are growing. The comb made by some species resembles a spiral staircase, and there are special cells for the pollen as well as for the honey; and here, for the first time, we find wax used to form the comb. Here also we find the three separate castes, the queen or queens, producing eggs; the working bees, or barren females; and the drones, or males. In some nests the wax is mixed with resin or gum, which makes it darker. The inside of the nest, like the inside of the hive of the honey-bee, is dark. Sufficient food, consisting of pollen and honey, is placed in each cell, and on this the egg is deposited by the queen bee or queen bees. Occasionally a bee leaves the nest, and apparently in many nests all three castes are reared in identical cells on a similar diet. There is no such specialization as that shown in the hive of the honey-bee. Another point of difference is that among the mosquito-bees the cells in which the larvae mature are sealed up. After the egg is laid there is no contact between the larva and the mother or the workers. The drone has not degenerated into the "waster" that he becomes in the hive of the honey-bee but takes part in cementing the wax for the cell walls. The entrance to the hive is guarded during the day by certain sentinels and is closed at night by a mixture of wax and gum.

When we come to the bumble or humble bee we find still further progress toward the state of things we find in the hive of the honey-bee. The bumble-bee has a sting but seldom uses it, and as the poison is weak the pain it inflicts is much less than that produced by the sting of the honey-bee. On the other hand, when once the bee has stung it can withdraw the sting and use it again. This the honey-bee cannot do; its stinging results in its death. The life of the bumble-bee is less orderly than that of the honey-bee. There is less of that irritating efficiency, and there is much more litter; after all what would life be without litter! Bumble-bees are found nearly everywhere in the world except in Africa and Australia, but they prefer a temperate climate. There are hundreds of species of this genus, and seventeen of these are found in Great Britain.

To describe the life of the bumble-bee we may begin with

the queen late in the summer. The nest is now dying down; in fact, the activity of the hive lasts only a few months, not all the year round, as does that of the stingless bee. Late in July or August the community begins to rear up queens. Once grown up the queen leaves the nest and hides in some cranny or among some débris. Here she is sought by the male. Once fertilized, the queen abandons the nest, which falls into a state of "death, damnation, and decay." She now seeks winter quarters and, having filled her crop with honey, she goes into retreat for eight or nine months, hiding high up in banks or in burrows under trees. At first she sleeps lightly and can be easily aroused. Later she sinks into a deep lethargy and appears to be dead. But as the spring advances she gradually resumes her activities. She emerges and begins to collect pollen. As the days lengthen her desire to start a colony becomes overwhelming and she seeks a home. She may find some burrow abandoned by a fieldmouse, which is commonly approached by a tunnel.

Having found her home, she flies backward and forward from it, gradually increasing the length of her trial flights. This she does so that she may find her way home after raiding the flowers for pollen and honey. She mixes the two, and in the centre of the nest constructs a small pillar of the resultant paste (Fig. 14), and on this she moulds a circular wall of wax. In this rough, irregularly-shaped cell she lays a batch of eggs, usually about a dozen, and seals them in with wax. She then broods like a hen over the cell and does not leave her offspring night or day except to gather food. But she has to provision the nest, and for this purpose she prepares a waxen spherical honey-pot, which may be as big as a thimble. This is a frail affair of thin, soft wax, but it is water-tight and is capable of lasting some weeks. Arriving at the entrance to the nest, the queen refreshes herself as she is passing in and out, and by night the honey-pot may be quite empty of its thin and watery contents. In about four days the larvae hatch out as whitish grubs and begin to feed upon the pollen bed upon which they have been lying. At first they feed upon any mixed pollen and honey provided by the queen. As they grow older they are individually and compulsorily fed.

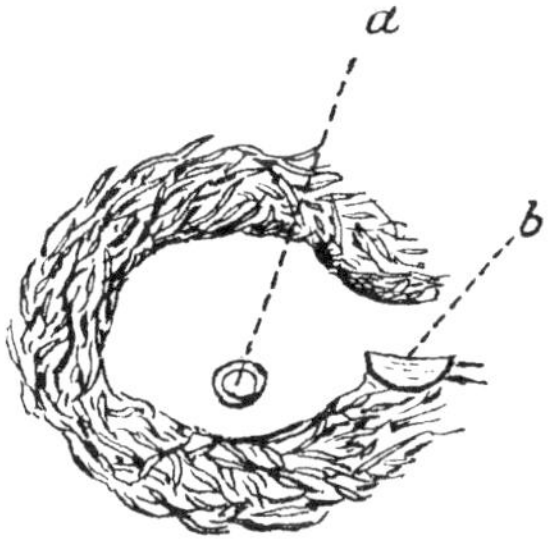

FIG. 14.—The beginning of a bumble-bee's nest, showing at *a* the pillar of pollen and honey on which the queen will deposit her first eggs, and at *b* the honey-pot.

In a week the grub-like larvae turn into chrysalids and spin about their bodies a thin, papery, but tough cocoon. The queen now removes what is left of the waxen cell, and the pale little cocoons stand on their ends like mummies. The outer rows are taller than those in the centre, and in the groove thus formed the queen lies brooding over the pupae, which hatch out on the eleventh day, when the complete female working bumble-bees step out into the darkness (Fig. 15). At first they are weak and tottery, yet they manage to make their way to the honey-pot and take a deep draught of the thin fluid before returning to safety beneath the body of the mother; but in two days they grow up and begin to help in the work of the nest. They start collecting pollen and honey as a store of food for the second and later broods of larvae, for the queen is now laying batches of eggs every few days. In fact, the second batch of larvae is ready for the attention of the lately hatched first batch. In the hive of the honey-bee the workers do not set about gathering food till they are two weeks old, but in the home of the bumble-bee this task is undertaken by the workers at the end of two days. A further distinction is that the honey-bee collects either nectar or pollen, but not both on one journey; and the pollen is usually uniform in colour, which indicates that it has been collected from one species of plants; but the bumble-bee during one flight brings back both nectar and pollen to the nest, and the pollen is obtained from dif-

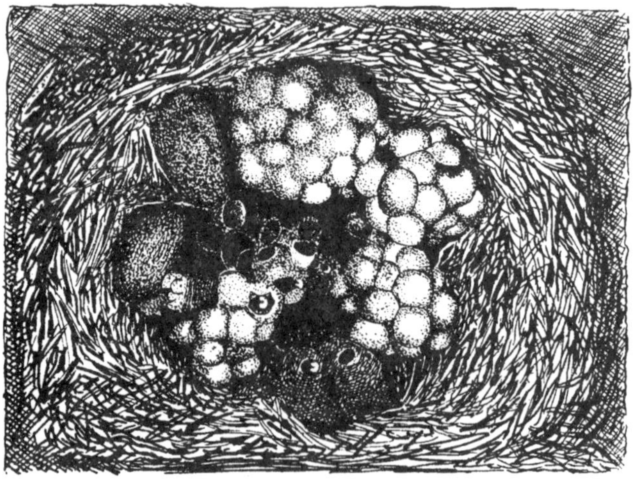

FIG. 15.—Comb of a bumble-bee, showing two honey-pots full of honey and two old cocoons stored with pollen. The irregular cells shown contain developing bees. Some of the cells have been opened and a young grub can be seen lying in the interior of the cell. (After Sladen.)

ferent sorts of plants, so that the thighs are streaked with white, lemon-yellow, orange, and bright-red pollen grains.

Should the first three or four batches of larvae hatch healthy and vigorous workers, the queen, who is now evidently tired, ceases to leave the hive and confines herself to laying eggs and helping with the necessary indoor work. When fully grown the workers cease to use the queen's honey-pot, which now falls into decay, and they store in the papery cocoons they have vacated the honey they have themselves collected, strengthening the edges with wax.

Some species of bumble-bees construct special honey-pots of their own, as many as twenty or more in a hive. These contain a very watery syrup, which is eaten up daily, but the honey in the cocoons is thicker and seems to be used to feed the younger queens. A few species make special receptacles for the pollen, which is mixed with honey. The comb made by these species is irregular and rough compared with that of the honey-bee, which shows mathematical rigidity. It is placed on the basal irregular waxen layer of vacated cocoons, on which also are placed cells containing larvae and pupae. Sometimes the whole comb may be covered by a waxen dome, but there is always room left for the bumble-bees to circulate. These bees have evidently an acute sense of smell, and human breath is particularly distasteful to them. They are almost as clever as honey-bees in their power of scenting out nectar and, owing to the length of the tongue, a bumble-bee can probe flowers to reach nectar that lies beyond the reach of the honey-bee. The bumble-bee fertilizes the honeysuckle, the horehound, and the red clover, whose introduction into New Zealand proved a failure until bumble-bees were brought in to fertilize it.

The hive of the bumble-bee is kept up for only three or four months. But the inmates are very busy; in fact, they work themselves to death. They begin foraging earlier in the morning than the honey-bee and they continue foraging till dusk. They spend the night in attending the young and brooding over the cocoons, for they never sleep. After laying from 200 to 400 eggs and slaving to bring up her progeny, the queen, as the season closes, begins to lay special eggs that are destined to turn into males and fertile females.

The cocoons for future queens are larger and may be readily recognized. But at present there is no evidence that the queen larva is fed on a special diet. Royalty seems to be inherent in the egg and not induced by special feeding. Unlike the useless and swaggering drone of the honey-bee the male bumble-bee leaves the hive and finds flowers for itself. It is no charge on the resources of the community. Several scores of males and queens are produced, and when hatched out they also leave the hive, are fertilized, and go into winter quarters. The queen ages rapidly; her hair drops off and she gradually ceases to lay eggs. As the new queens grow up on the rich and ample store of food provided in the hive the workers become listless. Flowers are becoming scarce, and one by one the bees grow torpid and drop asleep, and from this sleep there is no awakening.

The bumble-bee is certainly more human and less exasperating than the honey-bee. It has none of its monotonous perfection of organization. The queen has something of a mother in her. She is not reduced to a mere egg-laying apparatus, which lays eggs with the regularity and inevitableness of a recurring decimal. The bumble-bee queen broods over her young and nurses them with "a mother's tender care." The workers work as hard as do the honey-bees, but they are less self-conscious and less self-satisfied, and the drones at any rate have the grace to provide for themselves during their brief life. One has a feeling that one might appeal to the better instincts of a bumble-bee, but that it would be perfectly useless to make such an appeal to a honey-bee.

Now let us summarize the results of the research we have made to discover the steps in the evolution of the honey-bee, with its wonderful social system. The most primitive bee makes a small cell or nest in the ground (Fig. 16), packs it with pollen, usually mixed with honey, deposits an egg, covers the cell in, and leaves the young larva to eat up the food provided. The grub or larva then turns into a pupa, from which emerges the active adult insect, which makes its way out into the world. The next stage higher is shown in

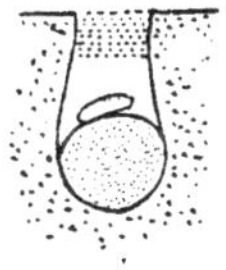

FIG. 16.

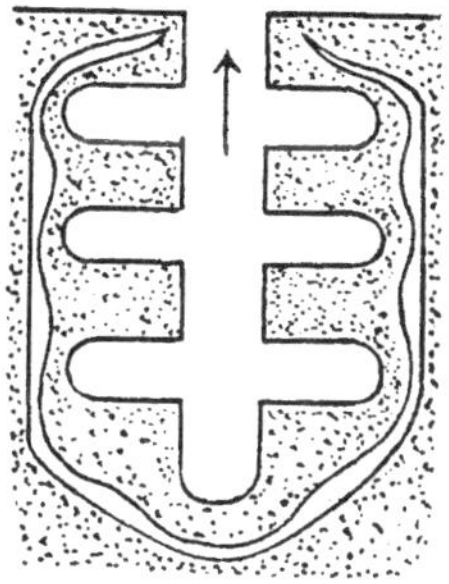

FIG. 17.

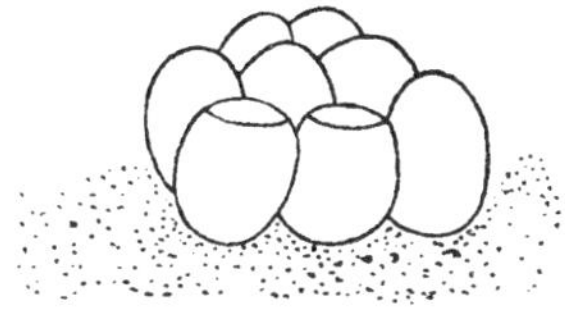

FIG. 19.

Fig. 18.

FIG. 16.—A single cell of a solitary bee made in the ground. The egg is deposited on a mass of pollen and honey and the cell is closed in.

FIG. 17.— A series of cells side by side but well separated from one another. These cells have a common passage indicated by the arrow, and the whole are surrounded by a common envelope. This is the first indication of a comb.

FIG. 18.—A row of cells of a solitary bee, such as the carpenter-bee. They are touching end to end.

FIG. 19.—A number of simple cells such as are found in the hive of the bumble-bee. They just touch one another, but have not really fused together, and there is no common wall separating them. The cells are really independent and are all made of pure wax.

Fig. 17. The cells are placed side by side or end to end, as are those of the leaf-cutting bee or the carpenter-bee; but each individual cell of both these bees is furnished with food and an egg and then left alone. The young bee does not receive any care or help from its mother. In the bees named the male may be smaller than the female and may not be differentiated into a lazy, idle drone.

In the next stage of progress we have a small colony, which inhabits a nest that has a common entrance, marked by an arrow in Fig. 17. In this stage the cells may be side by side, as in Fig. 11, or end to end, as in Fig. 18. In Fig. 17 the colony is surrounded by a specially protected case, such as we find in more complicated hives of bees and wasps. Finally, in Fig. 19, we find a number of cells side by side, which by pressure may become six-sided. Most primitive bees collect more pollen than honey and secrete no wax. Some bees make cells of leaves or of a substance that they secrete, which becomes papery; some carve cells out of wood; some cover their cells with a dome-like layer of mortar; and in many of these primitive nests the larvae spend months and months before hatching out.

When we reach the social bees—that is, the bees that live together in societies—we find that the most primitive are the mosquito-bees; but whether their communities are the product of a single queen or whether there is more than one egg-producer in their midst is not clear. Here we find, for the first time, special cells or collections of cells set apart for storing honey and other special cells set apart for storing pollen; and here, for the first time, we find wax, of which a comb is built up. This wax is a special secretion of the bee's body. Here again we find that the colonies have separated into queen, or queens; working bees, or females that do not lay eggs; and drones, or males. The fertile and unfertile females are reared on a similar diet. The larva is always sealed up, and once the egg is laid the young are deprived of a mother's care. And here again the drone takes part in the common activities of the hive.

The next stage toward the final product—that is, the honey-bee with its hive—is that of the bumble-bee. The hive of the bumble-bee, unlike that of the honey-bee, dies down during the winter, and the life instincts of the community are carried along in the body of the queen or queens, which retire into winter quarters early in the autumn. In the bumble-bee's hive as first formed there are no regular waxen cells, but the larvae grow up in an irregularly-shaped cell, which is sealed into a waxen covering. Over this the queen bee broods like a hen. Here also we have a specially prepared honey-pot situated near the entrance to the hive. This is not a modification of the ordinary cell, as in the comb of the honey-bee. In the bumble-bees the males take part in the work of the hive. The honeycomb is irregular and rough and may or may not be covered by a waxen dome. The queen bumble-bee lays only 200 to 400 eggs, which is a small number compared with the tens of thousands laid by the queen honey-bee. In the hive of the honey-bee we have true and exact hexagonal cells, each wall of which takes part in forming one side of the surrounding cells. Some of the cells are rather bigger, and these contain the heavy, overgrown drones. Others (but only a few) are still bigger and form irregular lumps of thick wax. Each of these big cells houses an egg, which is destined to become a queen. Other cells are set apart for the storage of pollen and still others for the storage of honey; but of course most of the cells that form the comb contain a single egg, which produces a grub or larva that receives hourly attention from the sterile workers, who act as foster mothers.

It has now been shown that there is a gradual development or evolution from a single pair of bees that make a single cell, isolated and self-contained, through a series of grades. The cells become more and more packed together till they reach the stage of the comb of a honey-bee, and there is developed a community of insects that rivals in complexity and in division of labour anything that we meet with in human communities. A clearer example of evolution could hardly be imagined—the gradual development from a simple primitive state of life to one of the highest complexity.

SIR ARTHUR EVERTT SHIPLEY

SALMONELLA

A few years ago while fishing for salmon on the Rogue River on the coast of Oregon, our Poodle -KoKo licked and chewed a little on the head of one of our salmon. The next day he had the appearance of having distemper. He was around the age of eight when this happened. His appearance grew worse and we took him to the vet. "He has Salmonella," was the doctors immediate reply, "and it looks pretty bad."

We left him there for a few days and he recovered. On picking him up the vet told us that we didn't have to worry about a recurrence because he was now immune to this disease. I can imagine that the Salmonella disease was a built in protection for the salmon as they headed for their spawning grounds hundreds of miles up the streams and rivers. Many creatures of the wilds patiently awaited the spring and fall runs of this very tasty fish on it's way up river. Undoubtedly this Salmonella was originally developed as some sort of protection and to deter certain animals from catching them on their journey. Any of the bear family were undoubtedly immune from birth as the salmon is a main-stay of their diet.

In my years in the wilderness areas I have never seen any of the cat family catch salmon. Neither a wolf or coyote have I seen in the act of catching or eating them. However one of their worst enemies, seals, follow them up-stream and entertain themselves while making a living catching salmon. Evidently the cats and dogs have never acquired an immunity and have learned their lesson.

BILL HUTH

THE LIFE OF THE BEE

"The bees have existed many thousands of years; we have watched them for ten or twelve lustres. And if it could even be proved that no change has occurred in the hive since we first opened it, should we have the right to conclude that nothing had changed before our first questioning glance? Do we not know that in the evolution of species a century is but as a drop of rain that is caught in the whirl of the river, and that millenaries glide as swiftly over the life of universal matter as single years over the history of a people?"—Maeterlinck's *The Life of the Bee.*

THE EVOLUTION OF ANTS

THE term "evolution" is used by biologists to cover one of the aspects of "development," which in turn merges into the universal phenomenon of "change." Most people regard change as a matter of course, merely noting its occurrence and its various forms and adapting themselves to it, whether it occurs in their own lives, in the lives of other human beings, or in the lives of animals and plants, but to reflective observers, during the past three thousand years, change has always seemed so extraordinary as to constitute the basis of philosophy or the occasion for philosophical speculation. Since the wonderful complexity and diversification of the world is due to change, and since to us the outstanding features in this diversification are human beings and other organisms, it is easy to see why the origin and meaning of change should have been sought and discussed so ardently, and for so many centuries.

There are three groups of facts with which even the most casual observer of the constantly changing organic world is familiar. First, he knows of the development of animals from eggs and of plants from seeds. Second, he sees the possibility and the usefulness of making a rough classification of animals and plants, and he notes that among the various kinds of animals and plants there are in nature certain forms (species), some of which are very similar, though distinct—such as the various kinds of oaks, pines, deer, and ducks—and even the superficial observer knows that these species, though they may be very constant in many of their characteristics, are nevertheless more or less variable in others. And third, everybody knows that many of our breeds of domesticated animals and plants have given rise and are still giving rise under human control to other breeds, some of which show great differences from their ancestors, such as those, for example, seen among our dogs, pigeons, roses, and grapes. The facts of the first and the third group we can observe directly; those of the second group, showing classification, require explanation.

The resemblances and the differences between the kinds of animals and plants might be accounted for in two ways: either these several kinds were created independently, simultaneously or successively, or they were derived by natural descent from common ancestors, in the same manner as the various breeds of domestic animals and plants were derived from their ancestral forms. The first explanation is supernatural and nongenetic; the second is natural and genetic. There is no question as to which of these explanations the scientist and the philosopher must prefer, for, as Joseph McCabe says, "no plea for the supernatural origin of anything is valid *so long as there is a possibility of a natural explanation of its origin.*"

The changes noted in the three groups of facts discriminated above all come under the head of "development" in its general sense, but those of the first group comprise the development of individual organisms, whereas those of the second and third comprise the development of races. The term "development," or "ontogeny," is now commonly used of individual development; the terms "evolution," or "phylogeny," of racial development. All reputable living biologists accept evolution either as proved or as so thoroughly substantiated as to be practically proved, but they differ as to the precise natural factors or conditions that have brought it about in any particular group of organisms. This point cannot be too strongly emphasized, because the discussions of biologists over the precise nature of the process of evolution are continually being misrepresented by ignorant or dishonest anti-evolutionists as confessions of disbelief in the occurrence of the process itself. There is surely nothing unusual about the discussions of evolutionary causes by biologists. Everybody now believes in what we call gravitation, but physicists, past and present, are by no means unanimous in their views of the precise nature or causes of gravitation. Those who are emotionally upset by the conception of a collateral generic relationship between men and the anthropoid apes may be reminded that although the difference in psychological and social behavior between the animals of these two groups is undoubtedly considerable, such behavior has exhibited great change even during historical time, and that the structural and functional differences between men and the anthropoids are trivial as compared with those which separate a frog's egg from an adult frog; yet this enormous gap is bridged by a continuous process that occurs under our very eyes. Endless confusion in the popular mind might be avoided if we could dissuade all journalists, politicians, teachers, and clerics from talking or writing on evolution till they had made an intensive first-hand study of the embryology of some animal or plant or a thorough investigation of some group of wild or domesticated animals or plants.

Now it is easy to prove genetic continuity within existing species and between the breeds or races of domesticated forms, but it is very difficult to prove genetic relationships between similar species of wild organisms, for the process of racial development or evolution is so exceedingly slow that even some slight structural changes may have required millions of years, or at any rate periods far too long to fall under the observation of a being so ephemeral as man. The proofs of this very long historical evolutionary process are therefore indirect; they derive their value from the convergent and mutually corroborative inferences drawn from studies made in widely different fields of the great science of biology. At least five of these fields furnish significant historical inferences—the study of fossil animals and plants (palaeontology), the comparative study of the development and structure of existing forms (morphology, or anatomy and embryology), the study of the present geographical distribution of plants and animals (chorology), the study of the classication of plants and animals (taxonomy), and the comparative study of the behavior of animals (ethology). Obviously, the study of extinct or fossil species is of the greatest value, but the record of some species is deplorably fragmentary and most of the specimens found are imperfectly preserved. Although, therefore, all positive palaeontological data are precious, the fact that we have not yet found connecting or intermediate forms at particular geological horizons may be of slight significance. Comparative morphology and its shorthand expression, classification, are of enormous value in determining the possible genetic relationships between species, both living and fossil, and the distribution of living species as compared with that of their fossil allies is of great historical significance. Finally, the study of the behavior of existing animals and of the dependence of behavior on the structure and function of particular organs enables us to draw inferences in regard to the actual modes of life of their allied extinct species. After these very general

FIG. 1.—Male ant embedded in amber.

Although this insect lived ages ago, the details of its structure are wonderfully preserved.

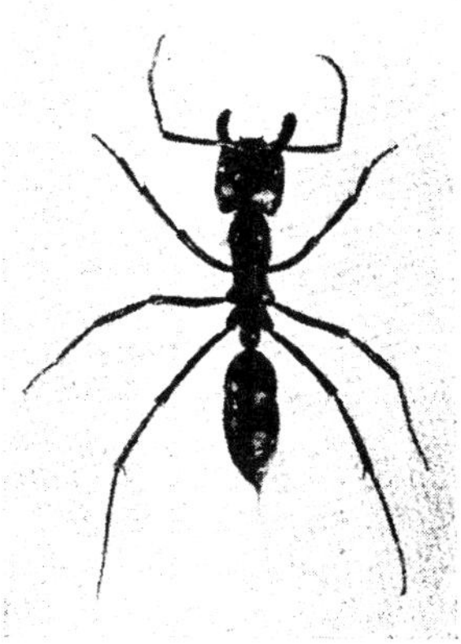

FIG. 3.—*Dinoponera grandis* of Brazil.

Worker, about natural size. (Photographed by C. T. Brues.)

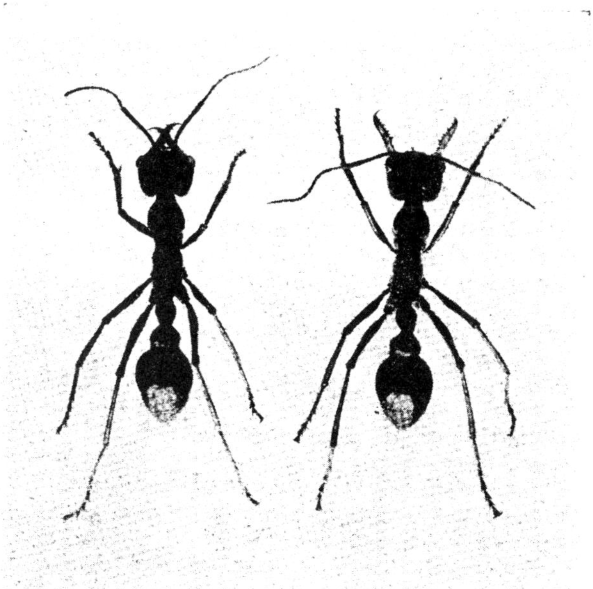

FIG. 2.—One of the famous bull-dog ants of Australia (*Myrmecia tarsata*). Of a beautiful deep-blue color. Mandibles yellow and tip of abdomen orange-red. (Photographed by C. T. Brues.)

statements we may turn to a study of the ants, which form one among a great many sources of inferences in support of evolution.

As a group, the ants are not so favorable for a study of evolution as their cousins the bees and wasps, because they constitute an unusually compact and homogeneous natural family and one which seems to have completed or nearly completed its evolution at an earlier date in geological time. This difference is indicated by the fact that all the six thousand or more known species, subspecies, and varieties of ants are eminently social, or live in organized colonies, whereas most of the wasps and bees are still solitary insects. There are also other reasons, which will be given later, for believing that the ants arose from a very ancient wasp-like stock and attained their present relatively high specialization a long time ago. We may now review some of the inferences derived from the study of their palaeontology, morphology, distribution, taxonomy, and ethology, which all agree in indicating not only that the ants have been subject to evolution but that this evolution has been of a particular character or pattern.

Many ants have been preserved in a fossil state in formations of Tertiary age, but none have yet been found in earlier formations. A small number of species have been found in Eocene deposits, which were laid down at the beginning of Tertiary time, but a much greater number have been collected from amber (a kind of resin) of Lower Oligocene age, found near the Baltic Sea, and from Miocene shales in Europe and in the United States, at Florissant, near Pike's Peak, Colo. Several species of ants have been found in Sicilian amber, which is also of Miocene age. I have studied no less than 10,000 specimens from the Baltic amber and at least 8,000 from the Florissant shales. Many of those in amber are exquisitely preserved (Fig. 1), having been enclosed in it much as insects are mounted in our laboratories in Canada balsam, so that they may easily be compared with existing ants, though the amber was formed millions of years ago. All this material, as well as that found in other formations and studied by others, shows that though the fossil ants, with a few doubtful exceptions, belong to extinct species, most of them belong to existing genera, and that none of the species is more primitive in structure and habits than many now existing. Indeed, many of them are quite as highly specialized as the most specialized existing forms. We are therefore unable to detect any significant evolution of the ants as a whole during the millions of years of Tertiary time, though many species have undoubtedly become extinct and others have arisen through relatively slight variations during that time and have given rise to the ants now living. We find, preserved in amber, even the larvae and pupae of certain ants, some of the plant lice which they tended, and a few characteristic ant guests (Paussidae) and parasites (mites). All this might seem to indicate that there has been no notable evolution of the group, but only a gradual extinction of species among a very considerable number that were suddenly created and distributed over the globe, but such a conclusion is unwarranted. We are bound to assume, on the contrary, that the significant vespoid or wasp-like forms among which the ants had their origin must have lived before Tertiary time—that is, during the Cretaceous period, or even during earlier Mesozoic time, which, unfortunately, is represented by few fossil insects, even of other groups. The only important conclusion we are at present justified in drawing is that the ants are a very old group of insects, which long ago attained essentially its present stage of evolution and has since been marking time or changing very slowly and imperceptibly. Probably the same was true of the ants of periods antedating the Tertiary, though there may have been in those periods occasional spells of acceleration and efflorescence of new forms.

When we carefully study the anatomy and development of the various species of ants we find that they are essentially wasps, and that they are closely allied to species of certain existing families of wasps, the Tiphiidae, Mutillidae, and Thynnidae. We must, indeed, suppose that the ancestors of these families produced also the ants, the Formicidae. But the members of these families, like most other wasps, are solitary, and, like most animals, possess only a single type of female; whereas among the ants each species presents two female phases, or castes, one of which, the "queen," is fertile and nearly always winged, and the other, the "worker," is always wingless and nearly always sterile. In only a few species of ants, and those highly parasitic species, do we find no worker caste. There is every reason to assume that in these species the worker has been lost or suppressed within comparatively recent time. We must therefore conclude that sexual trimorphism—that is, the presence in each species of three castes, male, fertile female, and sterile female, or worker, which were perfectly developed also in the known fossil ants of Tertiary time—was first established among the Mesozoic ancestors of the family Formicidae. A similar trimorphism has arisen independently among the social bees and social wasps, but it has evidently been of much more recent development, for among these insects the worker is much more like the fertile female and always has wings. Then, too, the differentiation of fertile and sterile females among certain tropical wasps is so feeble that the evolution of the two castes may be said to be still uncompleted.

When we arrange all the species of living and fossil ants

according to their structure we find that they fall into some seven subfamilies and that these may be most naturally regarded as seven large branches that arose from a single main trunk representing the most primitive and most wasp-like forms (Figs. 2 and 3). The existing species correspond to the green twigs and leaves at the tips of the branches of this "Stammbaum," or phylogenetic tree, and the new species that are discovered from time to time may be placed very naturally among their nearest allies according to this arborescent scheme. Now such an arrangement of the six thousand known Formicidae is the only one that will adequately represent the similarities or the relations of the forms, and the attempt to represent the morphological affinities of the species of any other group of organisms invariably produces the same kind of arrangement. This arrangement, moreover, would seem to admit only of a genetic or evolutionary interpretation.

It is, of course, impossible to give here any adequate account of the distribution of ants. With the exception of a few species that have been accidentally transported within recent times by man from one to another country, all ants are confined to rather narrow areas of the earth's surface, and their distribution agrees in general with that of other organisms, suggesting that the genera and species arose at different periods during geological time and then, with more or less modification, radiated to other regions, except as natural obstacles, such as large bodies of water or high mountain chains, may have prevented. The facts that most ants nest in the soil, that they avoid soil that is too constantly wet, that they are fond of warmth, and that they are abundant in certain arid regions suggest that they had their origin as a group on rather high continental areas during Mesozoic time. Many species, however, have since become adapted to life in dry deserts on the one hand and in moist, tropical jungles and rain forests on the other. Many of the ants in these jungles and forests, owing to the seasonal drenching of the soil, build their nests in trees or inhabit the pith cavities of twigs and branches. We also observe that regions of the globe like Australia, which are inhabited by the most primitive mammals and birds (duck-bills, echidnas, marsupials, emus, etc.), are also inhabited by the most primitive ants (bull-dog ants of the genera *Promyrmecia* and *Myrmecia*), whereas countries like Europe and North America, which have highly specialized mammalian and bird faunas, are similarly inhabited by highly specialized and dominant ant faunas, with which, however, are intermingled a small number of primitive forms, which were once widely distributed but are now rare and are in process of extinction. Such a distribution can be explained only on the theory of evolution and is in complete agreement with all we know about the geological history and morphology of other organisms.

Conclusions from a comparative study of the habits of ants, or ant behavior, which is necessarily restricted to living forms, agree closely with the conclusions reached in the fields mentioned. Although all ants are social, they exhibit different degrees of social organization. This diversity is shown in different degrees of division of labor in the colonies as coördinated with their size and in differences shown by their component individuals. Thus among the most primitive ants many of the colonies are very small and the fertile females and workers are much alike in size and structure, but in the most highly socialized species (Dorylinae, Formicinae, and Myrmicinae) the colonies may be very large and the workers may be unlike the females and may even exhibit a differentiation of the worker into secondary castes, major and minor workers (Fig. 4), or soldiers and workers proper. Along with this advance in diversity of form, there is a notable change in feeding habits, the primitive forms being purely carnivorous, like their ancient wasp-like ancestors, and the more advanced types having become increasingly vegetarian. The vegetable feeders are best developed in regions where competition for insect food is keenest—that is, in deserts, where insect food is scarce or limited to a short season, and in the tropical rain forests, where the ants must enter into close competition with many other predatory insects and with insectivorous reptiles, birds, and mammals. In the deserts of the world (in southwestern United States, Mexico, Sahara, South Africa, Central Australia) we find that two kinds of ants have become adapted to a vegetarian diet, the harvesting ants, which feed largely or exclusively on the seeds of plants, and the honey-ants, which store in the crops of a special caste of worker a sweet liquid ("honeydew") collected from plant-lice, scale-insects, and oak-galls. In the tropical and subtropical forests of the New World a peculiar tribe of ants (Attini) have acquired the habit of making mushroom gardens in which they grow fungi as food. The garden beds are made of pieces of leaves, which they cut from the trees, or from the collected excrement of caterpillars or other insects that feed on plant tissues (Figs. 5 and 6). Fully a hundred species of these attine ants are known, and some of the larger species are at times very injurious to the agriculturist and the horticulturist, because they use the leaves of cultivated plants (sugar-cane, orange trees, etc.) as material on which to grow their food-fungus. Many of the most highly specialized termites, or "white-ants," in the Old World tropics have independently developed a similar habit of growing fungi. Among these insects, however, the substratum of the fungus gardens consists of triturated wood, which has been passed through the intestines of the workers.

THE EVOLUTION OF ANTS

One of the most striking of the evolutionary habits of ants is social parasitism, which leads colonies of different species to live very near or actually with one another. One of these

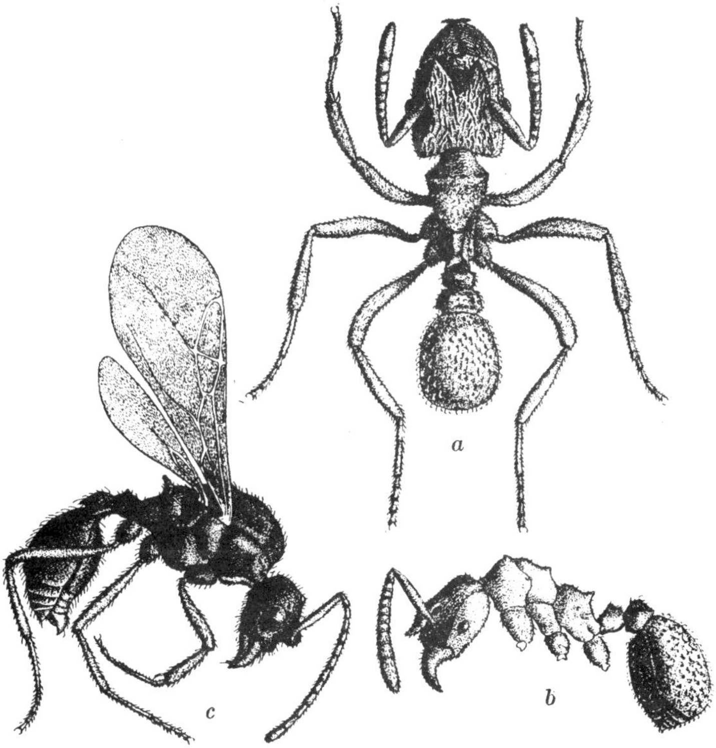

FIG. 5.—A small Texas ant (*Mycetosoritis hartmani*) that grows fungus. Considerably enlarged. *a,* worker, dorsal view; *b,* same in profile; *c,* male.

colonies exploits its neighbor, but the character of the exploitation varies. One species preys on the brood of another or enslaves it; another species uses its host merely for the purpose of bringing up its own brood; still another merely derives a certain protection from living near its neighbors. A careful study of these habits of ants has shown that they can be explained only as the results of a gradual and complicated evolution. This form of evolution has led to a peculiar degeneration of some parasitic species, which have, in fact, become abjectly dependent on the host for food, for the care of the brood, and for the construction of the nest, and some ants have even lost completely their worker caste. The strong conviction of naturalists that such parasites have been evolved from once independent organisms instead of having been created in their present dependent and degenerate form should be carefully weighed by all those who are busily attacking evolution in the name of religion and morality.

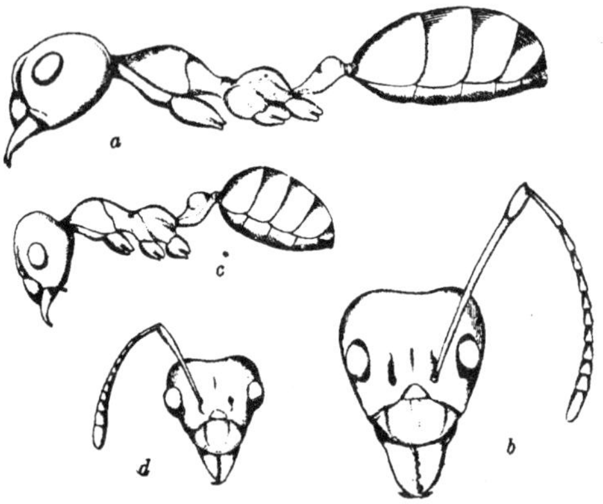

FIG. 7.— A weaver-ant (*Oecophylla longinoda*) of the Congo: *a*, major worker in profile, with legs removed; *b*, head of major worker from above; *c*, minor worker; *d*, head of minor worker.

That the activities of ants in response to particular environments have led to the development of highly specialized habits is shown also by many interesting examples of "convergent" or "parallel" evolution in species that are not closely related. One striking example is furnished by the tropical ants that inhabit silken nests on trees. These nests are really constructed by the young larvae, which their worker nurses use as weaving-shuttles. Throughout the East Indies, Northern Australia, India, and equatorial Africa the "tree ants" of the genus *Oecophylla* (Fig. 7) have attained great proficiency in the art of thus using their larvae for spinning adjacent leaves together (Fig. 8). A similar habit has also been acquired by certain species belonging to two other genera, *Polyrhachis,* in the tropics of the Old World, and *Camponotus,* in central and northern South America. The nests made by *C. senex* and *C. formiciformis* in the forests of British Guiana, Panama, and Guatemala are extraordinarily like those made by *Oecophylla longinoda* in the forests of the Congo and by *Oecophylla smaragdina* in the jungles of India. The structure as well as the behaviour of remotely related species of ants has been similarly modified by convergent or parallel evolution in response to identical environment. A fine example is furnished by certain ants in which the head is cylindrical, constructed like the cork of a bottle, with a hard, roughened, truncated anterior surface, and used for closing the circular orifice of the nest, which leads to galleries excavated in sound wood or in hard soil. Species of at least four different genera in different parts of the world (*Camponotus, Pheidole, Crematogaster,* and *Epopostruma*) exhibit this identical form of head. A similar modification of the head is seen in a number of worms, bees, beetles, toads, and tree-frogs; and in certain spiders, beetles, caterpillars, snakes, and armadillos the posterior end of the body is similarly modified for use as a barricade for closing the burrows in which they live and thus preventing the entrance of enemies.

A different modification is seen in certain ants that live in the narrow pith-cavities of the twigs and smaller branches of tropical shrubs and trees. In these insects the whole body becomes very long and slender, or even thread-like, or filiform. This singular modification, too, is seen in several different genera in the tropics of both hemispheres. Finally, attention may be called to the development of a peculiar beard, consisting of long, forward-sweeping hairs on the lower surface of the head in several unrelated genera of desert ants. The hairs are rather stiff and form a kind of crate or basket, in which the ants carry up the dust or sand that they loosen while they are excavating their burrows. To account for all these exquisitely adapted forms or features there are only two hypotheses: either they have been developed gradually, in response to the environment in which the insects have long been living, or they were created at the same time as their possessors by a being having a prevision of their ultimate function. If the latter hypothesis is accepted we can only marvel at the Creator's meticulous solicitude for the welfare of ants and His failure to provide adequate prophylactic measures against the many common diseases and calamities that have for thousands of years decimated the paragons of His creation.

Of course, the conclusions we have reached in regard to evolution among ants, though based on many more observations than those briefly cited here, relate nevertheless to a very small part of the animal kingdom. But during the last sixty years essentially the same conclusions have been reached by hundreds of other students, each of whom has investigated some particular group of organisms; and the combined labors of all these workers may be said to cover the whole extent of the plant and animal kingdoms, man included. Are we to suppose that this conclusion, unanimously reached by so many men who have devoted their lives to minute and conscientious observation and experiment, is the result of some marvelous unanimous hallucination, and that the truth lies with those who have given little or no study to the organic world, but have accepted blindly the doctrine of special creation fostered by equally unobservant ancient Hebrews, Babylonians, and mediaeval priests? In this matter, however, mere appeal to authority, either scientific or theologic, is unnecessary. The innumerable unequivocal facts that have convinced all competent biologists of the reality of the evolutionary process are recorded in thousands of volumes, which are open to the perusal of all who cannot find opportunity to make independent observations of their own.

WILLIAM MORTON WHEELER

ORGANIZATION

"The Anthropoid apes no doubt approach nearer to man in bodily structure than do any other animals; but when we consider the habits of ants, their social organization, their large communities and elaborate habitations, their roadways, their possession of domestic animals, and even, in some cases, of slaves, it must be admitted that they have a fair claim to rank next to man in the scale of intelligence."—Sir John Lubbock.

THE SOUL SPIRIT

The Spirit which lives in both you and me and all creation is an unending continuation of the Ether Spirit of life. You live with It every day, and sleep with It every night. You can't leave It and all creation exists with It as a part of them. This Ether Spirit is in contact with your Soul Spirit continually. The Spirit which belongs to you is the Spirit of your Soul. It makes no difference which shape your Soul takes on, your Soul Spirit remains with your Soul for all Eternity. Everything within your Soul has been reborn millions and millions of times, while your Sole has been evolving steadily trying to achieve what most of us have in mind as heaven on earth.

Bear in mind that the Spirit of your Soul is that voice with which all of creation converses in themselves. It is a beings direct contact with the Ether Spirit of the Universe, from which this magic life started.

Alpha and Omega are only in the mind and came to existence as a crutch to attach our own unexplained being; just as mankind, on their own, try to measure time. The unknown Ether Spirit has been forever, while thousands of gods have come and gone yet only those who tell about them have "seen" these idols.

It is unnecessary, or needless to worship in this Spiritual World: The Spirit must not be a thing of fear, for without you who are a part of a creation that the Ether Spirit cannot exist without. The Ether Spirit must have creation to Be.

BILL HUTH

THE EVOLUTION OF THE HORSE AND THE ELEPHANT

THE horse and the elephant are so well known and their characteristic features are so striking that a study of the changes which have taken place in their ancestors to bring them to their present forms should be of general interest. The horse and his associates, unlike other animals, has on each foot only a single toe—the hoof—and the elephant is unique in possessing that wonderful organ, the trunk, which is adapted to so many uses. In our study of the evolution of these animals we shall have to turn to the geologist for the evidence, which consists of bones entombed in beds of sand and clay, most of them now hardened to rock, laid down in different parts of the world during what is called the Tertiary period (see the accompanying geologic time table) and part of the succeeding Quaternary period, in which we are now living. The order of succession of the animals whose forms are thus revealed must be determined by the order of the deposition of the beds in which the bones are found. In a series of such beds the one at the bottom was laid down first and the overlying beds were laid down in the order in which they appear, one above another. The bones found in these beds belonged to animals that lived and died about the time the beds were formed. Each bone found occupied a certain known position in the skeleton of the animal, had certain distinctive features and performed certain definite, well-known functions.

The Horse

The work of tracing the development of the horse is relatively easy, especially the forms of the horse that lived in America, for we have a very extensive series of fossil remains of American horses, taken from beds that are piled in succession one upon another (Fig. 1). More than 200 different kinds of American horses have been discriminated, and in addition to these about 30 kinds have been found in Europe and nearly as many more in South America, Asia, and Africa. Some beds have yielded thousands of teeth and jaws, some have yielded other parts of the bony frame, and most of the types of horses are represented by complete skeletons.

GEOLOGICAL TABLE SHOWING EVOLUTION OF THE HORSE

Period	*Epoch*	*Length of epoch in millions of years*	*Millions of years ago*	*Kinds of horses*
Quaternary	Recent Pleistocene		1	Equus
Tertiary	Pliocene	8	9	Plesihippus, Hipparion Pliohippus, Hippidium Protohippus
	Miocene	13	22	Merychippus Parahippus Hypohippus Anchitherium
	Oligocene	13	35	Miohippus Mesohippus
	Eocene	20	55	Epihippus Orohippus Eohippus

The genus that includes the modern horse (*Equus*) is represented today by the domestic horse (of which there are many breeds), the wild horses of Mongolia, the half asses (the kiang and the onager), the asses, and the zebras. The genus includes eight to twenty species of living animals, the number discriminated depending upon the judgment or the fancy of the naturalist making the classification. Most of the later fossil horses are so nearly like the living horses that they have all been placed in the genus *Equus*. Any one who sees restorations of these animals at once calls them horses, zebras, asses, though some are much smaller than the living animals of these kinds.

North America was an early home of the horse, whose remains have been found in deposits in Wyoming that were laid down in Eocene time. (See the geological table.) At that time the climate of North America was warmer than it is now and Alaska was linked to Asia by land over which horses migrated. The Eocene lignite beds and gypsum deposits of France contain abundant bones of horses, and bones are found also in England, which was then connected with the Continent.

The earliest horses whose remains are found in America are the Eocene forms known as *Eohippus* (the "dawn horse"), some of which stood only about a foot high at the shoulder. The fore foot had four toes, the hind foot three toes, but each showed a vestige of an additional toe. The teeth were simple and short. Three kinds of Eocene horses have been distinguished, called *Eohippus, Orohippus,* and *Epihippus.* These horses appear to have lived in Western North America and in England at nearly the same time. Some of them appear to have inhabited either park-like openings in forests or the forests themselves.

By the end of Eocene time or a little later all the European horses seem to have disappeared, for the deposits laid down in Europe about that time contain no bones of horses. In

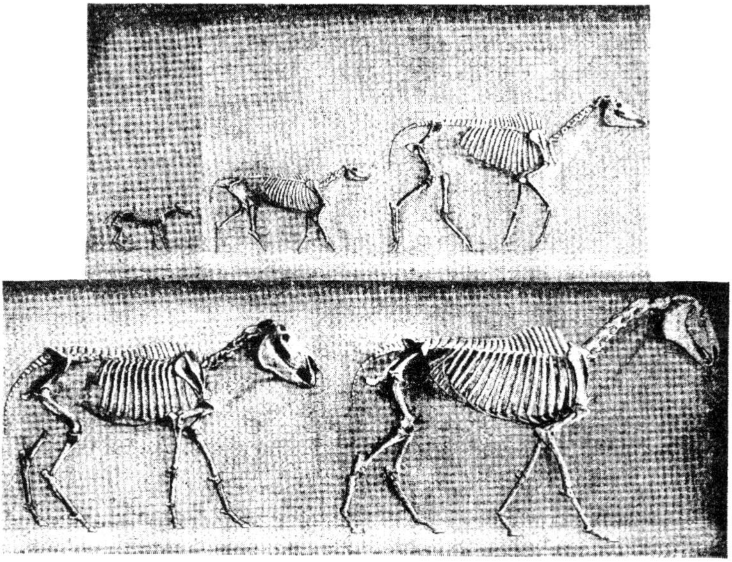

Fig. 1.—Evolution of the horse. Stages passed through from the earliest four-toed *Eohippus* (upper left) to the three-toed *Mesohippus* (upper middle) to a form with the side toes reduced (*Merychippus*, upper right) to the small, clumsy, one-toed *Equus scotti* (lower left) to the modern horse (lower right).

Reproduced from photographs by W. E. Corbin of fossils found in successive layers of rocks and now preserved in the Museum at Amherst College.

America, however, they flourished, and they continued to multiply and to grow larger during the following epoch, the Oligocene.

The Oligocene horses were larger than their predecessors, and each of their feet bore three toes. This three-toed horse is called *Mesohippus.* It appears to have been confined to the American continent, a fact suggesting the temporary severance of land connection between North America and Asia. The later Oligocene horses (*Miohippus*) were larger and had longer teeth and smaller side toes than the earlier forms.

During the Miocene epoch, which followed the Oligocene, Western North America was inhabited by many kinds of horses, among them one that has been called the "forest horse." The Miocene horses were also three toed and had low-crowned teeth. The forest horses spread from America to Asia and Europe, where their remains are found; but they appear to have died out in America in mid-Miocene time and later in Europe. The main line of the horses, however, continued to exist and underwent great changes, all originating in America. These changes appear to have been determined by environment. The teeth became longer and harder to adapt them better to grazing; the feet, which in the earlier horses were first five-toed, then four-toed, and then three-toed, advanced toward a single-toed form, the side toes becoming useless (Fig. 2). These changes indicate growing adaptation to life on grassy plains. The grass of these plains is harsher than that in or near forests, containing more silica, and horses that feed on it must have hard teeth. A hard, small hoof is also peculiar to plains horses, as well as long legs, for the horse must be able to escape from enemies, such as wolves and other carnivorous animals.

In their adaptation to life on plains in Miocene time the horses differentiated into three groups, the types of which are known as *Pliohippus, Protohippus,* and *Hipparion,* each probably representing adaptation to life on a certain type of plain—the grassy plain, the brushy plain, and the desert plain.

At the beginning of the next geological epoch, the Pliocene, the three types just named were living in America.

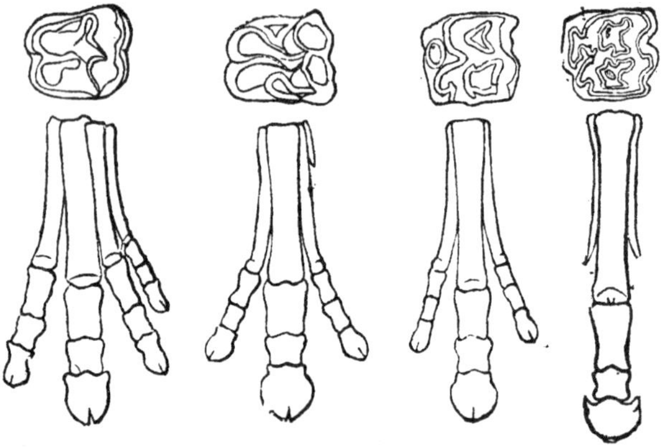

FIG. 2.—Principal stages in the evolution of the teeth and the fore foot of the horse. Showing the increase in the complexity of the grinding teeth and the gradual loss of toes on the front foot.

1. Four-toed horse (*Eohippus*). Eocene epoch.
2. Early three-toed horse (*Mesohippus*). Oligocene epoch.
3. Later three-toed horse (*Merychippus*). Miocene epoch.
4. One-toed horse (*Equus*). Pleistocene and Recent epochs.

During this epoch America was again united to Asia by a stretch of land that extended across Bering Sea. Over this land horses migrated from America to Asia and from Asia to Europe, where they became abundant and were differentiated into several species. Horses also found their way from North America to South America across the Isthmus of Panama, then recently emerged.

The Pliocene horses were all plains horses of the three

groups named. They were all three toed, but the side toes in most species were so small that they did not touch the ground. *Pliohippus,* which lived only in North America, had the smallest side toes; *Hipparion,* which lived in North America and in Europe, had the largest side toes and had teeth of a distinctive pattern.

In the Pleistocene epoch, which followed the Pliocene and which included the Ice Age, horses were abundant in both America and Europe. These horses, which were presumably the descendants of *Pliohippus,* are so nearly like the horses now living that they have all been placed in the genus *Equus.* Pleistocene deposits found in all parts of the United States, from California through Texas to Florida and northward to Nebraska and Pennsylvania, have yielded remains of horses, some smaller than even the smallest living pony, some as large as any living horse, and one species (*Equus giganteus*) the largest horse known.

The Pleistocene epoch is the period of maximum development of the horses in number, size, and variety. During this epoch horses made their way from America by way of Alaska and an isthmus across Bering Sea to Asia and Europe. Remains of Pleistocene horses are found in Alaska. The ice sheet of the glacial age, although it covered northeastern America, did not extend west of the Rocky Mountains, so that Alaska was then temperate enough to permit horses to live there. During this epoch horses made their way also to South America over the Isthmus of Panama and spread as far south as Argentina.

The most notable evolutionary changes in the horses consist of an increase in size, changes in the size and structure of the teeth, and, most conspicuous of all, changes in the form of the foot. The stock from which the horse was derived was probably five-toed, the foot conforming in its general pattern to the common mammalian foot, but the stress for speed appears to have centered on the third toe, leading to the elimination of four of the toes, resulting in a one-toed, swift-running animal, as the geological record shows. Relics of two of the last toes to disappear are seen in the splint bones of the modern horse.

The geological record reveals to us, of course, only the bony parts of the numerous horses whose remains have been preserved in the rocks and discovered. The differences in these parts have enabled us to discriminate many species, but if we knew the differences in mane and in tail and in colour we might increase greatly the number of species. We do know enough to assure us that a continuous series of horse-like forms inhabited the earth for ages and that Western North America was the principal scene of their remarkable development. All the American horses, however, finally became extinct from some cause or causes not yet discovered, perhaps a parasitic or a contagious disease. Thus the horse, which lived and developed for more than forty million years in America, died out on its native soil; but it survived in the Old World, though in smaller numbers and in forms less varied than it had in America in the Ice Age.

When America was discovered and explored by Europeans there were no horses in the country. Although some horses escaped from the Spanish conquerors and became wild, both in North America and South America, our domestic horse is a descendant of European breeds, which are numerous and extremely diverse in size as well as in other features.

In early historical time the domesticated horse was used to draw chariots and as a riding animal. From the earliest stages of its domestication it was highly prized. The horses of one country were traded for those of another, or were captured, and great care was taken in their breeding. Some of the various breeds differ more widely than some wild species, and the changes made under domestication have been far more rapid than those which occur among horses in a natural state. Among wild horses the changes that occur in a thousand years, or even in a million years, may be slight, but by successive changes from age to age we proceed from the tiny four-toed horse of Eocene time to the large single-toed horse of to-day. Every year has revealed more and more intermediate forms, until the fact of the gradual evolution of the horse is now recognized by all biologists.

THE APPALOOSA HORSE

As described in the American Heritage Dictionary, the Appaloosa horse is a breed of saddle horse developed in the northwest section of North America, characteristically having a spotted rump. (origin unknown.)

I have spent many years raising and training horses for riding and packing. During this time I had all sorts of breeds and colors. I have always been very fond of Appaloosas because of the coloring and the outstanding spots which cover their bodies. Some were completely covered and some only had a few spots on their bodies.

This peculiar part of the Appaloosa strain were their mane and tail and the wrinkled skin which surrounded their eyes. First of all the mane, like the tail, has sparse hair, kind of like they had been losing it. It didn't fall out but just seemed to be that way for their entire life. Most peculiar though were their eyes. When I first became aware of the problem the Appaloosa had with their eyes I was taking a party of fishermen on a pack trip up Slough Creek in the north east corner of Yellowstone National Park. One of the guides was leading the party and I was picking up the rear and following one of the fishermen, who was riding an Appaloosa horse I owned by the name of Chief. There was a single pipe gate across the trail and it was necessary for the horses and riders to go around it. The rider on Chief never made an effort to turn him and he walked right into the gate. No one got hurt but this was the first time I ever saw a horse do this. The next time it happened, my son, Chris, was riding him. We were coming out of the Beartooth wilderness outside of the park and he was following me down a pretty steep trail. All at once I hear him call to me, "hey Dad, look at this." I pulled my horse up and looked back. Chief had walked out on the edge of a boulder and Chris had stopped him there; just ahead of them was a drop off for over one hundred feet which they would have gone off of if Chris hadn't pulled Chief up and stopped him. I learned later that his problem was defined as night blindness, but when further along in development it wasn't limited to night time riding. Chief got worse and worse and we finally had to put him down.

This particular strain, Appaloosa, seems to have come from the Nez Perce Indians who live in the same territory as the Appaloosa horse. It's a common saying about the Nez Perce losing the war with the white man because they were riding Appaloosa horses.

BILL HUTH

THE EVOLUTION OF THE ELEPHANT

The elephant is not only the largest living land animal but the most peculiarly built. Its wonderful prehensile trunk and its long, heavy tusks are its most striking features. (See Fig. 3.) It has relatively short and straight legs, a short body, and a very short neck, so that its head is carried high above the ground; its lower jaw is short, it has no front teeth, and only one grinding or cheek tooth in each jaw. Each grinding tooth is composed of 17 to 25 plates and weighs 15 to 20 pounds. Elephants' teeth are so large and hard that they form fossils which are easily recognized. Fossil teeth and bones of elephants and of their relatives, the mastodons, have been found in North America, South America, and Europe, countries in which elephants no longer live, as well as in Asia and Africa. The elephants appear to have always lived in dense forests and to have fed on the vegetation they afforded.

Only two species of elephants are living to-day, the African and the Indian elephant, but in the Ice Age, which occurred in the Pleistocene epoch, there were more than twenty species. In the Pliocene and Miocene epochs there were few true elephants but many mastodons, and they inhabited all the continents except South America and Australia. In the Oligocene and Eocene epochs there were no true elephants, and the mastodons appear to have lived only in Africa.

The earliest known member of the elephant family is a tiny form, just over two feet high, whose remains have been found near the Fayum Oasis, in Egypt. It is of upper Eocene age, and is a mastodon known as *Moeritherium.* This primitive mastodon may have given rise to the true elephants, although its neck is short and heavy, its feet are short and compact, its head is of normal length, and it has nearly a full set of teeth. The second incisor in the upper jaw, however, is large, and the third incisor and the canine tooth are small and appear to be on their way to being lost. It also has incisors in the lower jaw, the second one large, but no third incisor or canine tooth. In each jaw there are six low-crowned grinding teeth. This form seems to have lived near rivers and ponds and to have subsisted on soft vegetation, the large upper and lower incisors suggesting that they were used to dig up bulbs and roots. Remains of mastodons in this stage of development are found only in Egyptian beds.

In the Oligocene beds, which lie just above the Eocene, the remains of another elephant-like form are found. This form which is called *Palaeomastodon,* has the second incisor of both the upper and lower jaws considerably enlarged. Of the other teeth, the first and third incisors and the canine of the upper jaw have disappeared, as has also the first incisor of the lower jaw; all that is left of the front teeth in either jaw is the second incisor. These incisors of the lower jaws are flattened and the jaw is elongated, so that in spite of the fact that the neck has shortened the mouth still reaches the ground. The two incisors of the upper jaws have greatly enlarged and spread to either side. Between these upper tusks lay the upper lip, prolonged enough to reach to the end of the lower jaw and to manipulate the food dug up with the lower incisors. *Palaeomastodon* was about four and a half feet high and seems to have improved or specialized in the habit of digging roots and fleshy vegetables. In these same Oligocene beds of Egypt we find still another and more progressive form, *Phiomia,* which may well be called the "long-jawed mastodon," for the lower jaw is still longer and the neck is still shorter.

At the beginning of Miocene time there was in Europe a group of long-jawed mastodons closely related to *Phiomia.* They migrated from Africa to Europe and increased in size until they were about eight feet high. The lower jaw was as much as six feet long. After reaching Europe these long-jawed Mastodons spread over the continent and migrated to Asia and finally to North America across land that then connected the continents. In Pliocene time this form culminated in *Trilophodon giganteus,* which was almost as large as the later mammoths. In most animals the neck elongates as they increase in size, so that the mouth can be brought to the ground for feeding or drinking, but in *Trilophodon* the neck steadily shortened, and the necessity of reaching the ground has been met by elongating the jaws. The two large shovel-like teeth of the lower jaws indicate that these large forms were still digging roots and fleshy bulbs for food. The upper jaw is not so long, but the two upper tusks are long enough nearly to touch the ground and probably aided the lower tusks in digging and pushing aside the earth. The upper lip must have been correspondingly long.

This is a critical time in the history of the elephants. The dinotheres, mastodons, mammoths, and elephants of later time all seem to have gone through this long-jawed stage. In the Miocene epoch some of the long-jawed mastodons changed from the habit of digging to that of browsing on leaves and twigs. The shovel-teeth were no longer needed. The long lower jaws were no more an advantage. So we come to a series of elephants in which the lower jaw is shortening. The necessity for reaching the ground, at least for water, still remains, and for this purpose the long upper lip is used, but being no longer supported it becomes pendant, a proboscis. These new forms with a shortened lower jaw are the mastodons. There are many intermediate stages, such as *Mastodon longirostris* of the lower Pliocene of Germany, in which the lower jaw is still of considerable length. *Mastodon atticus* from Pikerni, in Greece, has a long chin, and while young has incisors in the lower jaw. Throughout the Pliocene there are several species of mastodons in Europe and Asia. By Pleistocene time they had reached North America, where they flourished throughout the Ice-Age and for a short time afterward. Among even the American mastodons there is occasionally found one which has vestiges of tusks in the lower jaw, like the one at Amherst College, which has tusks nine inches long.

The upper tusks, after they were no longer used for digging, did not disappear, as would be expected, but instead turned upward and forward, increasing in size, so that those of a well-grown adult are usually seven to eight feet long, and a single tusk weighs over a hundred pounds. Projecting so far in front of the animal they may be of some use in pushing or lifting, perhaps in fighting; but these uses can hardly compensate for the inconvenience of such projections, or for the effort involved in carrying so much weight out in front of the head. Like the antlers of the elk or moose, they seem to be overdeveloped structures, and were probably among the features that caused the extermination of these great beasts. Though the mastodons flourished throughout the Pleistocene epoch and for a short time thereafter, they all died out 15,000 to 20,000 years ago. They had attained a height of about ten feet, which is about the size of most of the larger living elephants. They were, however, shorter legged and more massive in build than the living elephant.

Though the jaw was shortened, the mastodons never developed complex grinding teeth, but had five rather simple grinders in each jaw. Apparently they remained browsers to the end of their days.

While the mastodons were developing their upper tusks and losing their lower ones, there arose a group of elephants which lost the upper tusks and retained the lower ones, though the lower jaw had shortened so much that it could no longer reach the ground. These animals were the dinotheres, which arose in early Miocene time from long-jawed mastodons and increased to a maximum size of but little less than that of the mastodons themselves. By late Miocene time they reached their height of development, only to die out in early Pliocene time. In the dinotheres the lower tusks were not simply retained but were enlarged and recurved. It is hard to guess the use to which such tusks could be put. The back teeth of the dinothere resemble those of the mastodon, and probably the animal had a proboscis, similar to that of the elephant, else it would have had no means of reaching the ground to drink.

In late Pliocene time, while the typical mastodons were browsing on leaves and twigs, some of the group began to feed on grass. As already suggested, grass carries considerable silica in its stems and leaves, so that animals which feed on it must have hard teeth. The first change taken to harden the teeth was to increase the number of cross ridges from three or four to six or eight, and then to increase the height of each ridge. Forms that attained this stage of development are known as stegodons. They were short-jawed and lived in Asia in Pliocene time. They were abundant both in numbers and in species. Some of them had enormous tusks, one tusk in the British Museum being nine feet and nine inches long.

The cross ridges of the teeth continued to develop in number and in height, and to supplement them, the cement, which usually is found only around the roots, worked its way up around the outside and into the valleys between the ridges, until the high ridges were welded together. Such teeth may show from ten to twenty-seven ridges and may reach a height of eight to twelve inches. From the time the valleys are filled with cement these forms are known as true elephants, mammoths if extinct, elephants if living. This increase in the size of the tooth, coming at the same time as the shortening of the jaw, has caused a curious manner of succession in the teeth of elephants. The first (rather small) grinding tooth comes into place soon after birth. It is used and worn for two or three years. Then the second grinder comes up behind it, crowds it out toward the front and takes its place. In a similar manner, one after another, the rest of the grinding teeth come in, crowding out the predecessor, until in about the fifteenth year the last molar, the largest one, comes into position, and this one functions for the rest of the elephant's life, 200 years or so. These true elephants are grazing forms. They flourished in Pleistocene time all over the world except in Australia. In North America, along the ice front, roamed the woolly mammoth, *Elephas primigenius,* mostly about nine feet high, the same mammoth that roamed in northern Europe and Asia and that has been preserved for us, frozen in the ice, in Siberia and Alaska. In southeastern North America there were the Columbian and Jeffersonian mammoths, ranging from eleven to twelve feet in height. In the Southwest lived the imperial mammoth, the greatest of all the elephants, measuring in height thirteen and a half feet. Just after the Ice Age all these animals died out, just why it is hard to say; perhaps by some contagious disease.

Of the several Pleistocene mammoths in Asia, only one has survived, the Indian elephant. The Pleistocene elephants of Africa are less well known, but two species are still living on that continent, the African elephant, some forms eleven feet high, and a dwarf form from the Congo, which is but four to five feet high. With the changes which took place in Europe after the disappearance of the ice sheet, some land areas became separated from the mainland, and the elephants on these new islands were restricted in their wandering and breeding. On Malta, for instance, there were developed two dwarf forms, related to the elephants of the mainland but consistently small—*Elephas melitensis,* some five feet high, and the tiniest of all elephants, *Elephas falconi,* but three feet high. Sicily, Cyprus, and Crete also had dwarf varieties.

The successive types that have been thus briefly described form a regular series, illustrated in Fig. 4. On looking at this figure we find it impossible to resist the conclusion that we have here the stages in the evolution of existing elephants —that these animals have come into existence by a series of gradual changes. Little swamp-dwellers with numerous simple teeth capable of crunching succulent aquatic vegetation become adapted, step by step, to life in a forest. The limbs were converted into stout pillars, to support the increasing bulk of the body and to stamp down small plants, and the toes were fused together into an insensitive mass, practically unpierceable by spines or thorns. The snout was drawn out into a muscular, flexible proboscis, capable, on the one hand, of gathering up ground vegetation and, on the other, of taking toll from the foliage of trees.

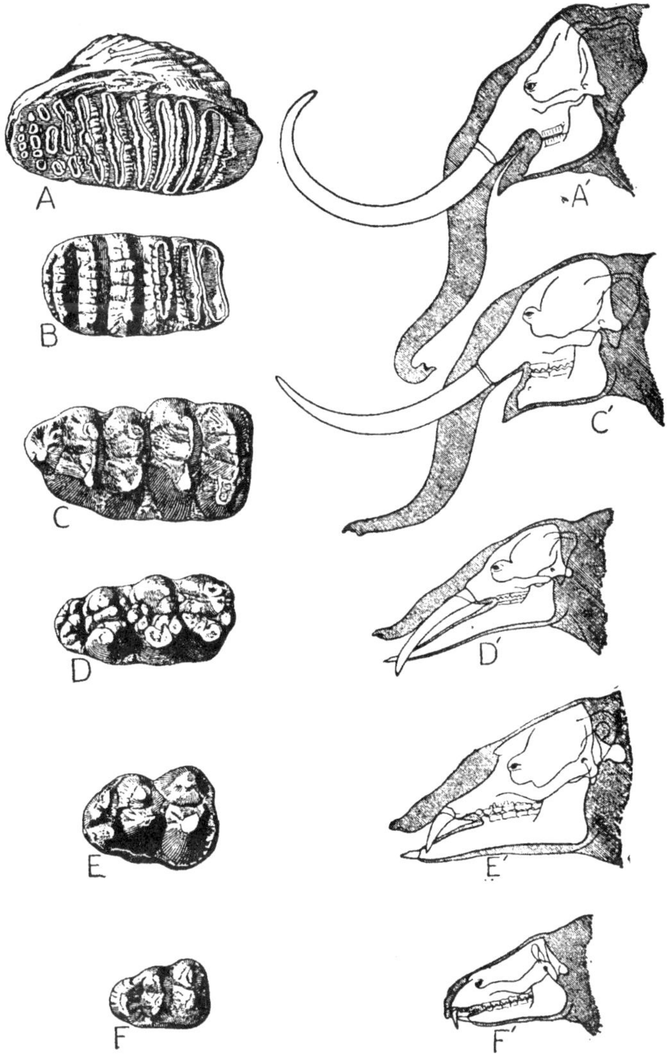

FIG. 4.—Evolution of head and molar teeth of mastodons and elephants. A, A′, *Elephas,* Pleistocene; B, *Stegodon,* Pliocene; C, C′, *Mastodon,* Pleistocene; D. D′, *Trilophodon,* Miocene; E. E′, *Palaeomastodon,* Oligocene; F, F′, *Moeritherium,* Eocene. (After Lull.)

The front teeth were reduced in number, those which remained becoming tusks, found in the lower as well as the upper jaw and first used for digging. Ultimately the lower tusks were suppressed, the lower jaw was shortened, and the upper tusks became enormously developed, to serve as weapons and to present a firm surface across which the trunk could break torn-off branches into pieces. The grinding teeth also became fewer and at the same time larger and more complex, to constitute an effective milling apparatus for crushing tough vegetable food. In an adult Indian or African elephant only four of these enormous teeth are in place at the same time, and each of them consists of three substances of different degrees of hardness, which wear unequally, so that the crown is always kept rough, for if smooth it would not be an efficient grinding surface. The densest of the three substances, the enamel, projects in a series of transverse ridges, narrower and more numerous in the Indian species than in its African cousin.

The head has necessarily become large and heavy, having to support the massive trunk and huge teeth (tusks and grinders) and to give attachment to the complicated muscles of the trunk and the large muscles by which the lower jaw is moved up and down in chewing. A long neck is obviously incompatible with so large a head, though the weight of this is less than might have been expected, for the unusually thick wall of the skull is not solid but is traversed by complicated air spaces.

The story of the Elephant, thus briefly outlined, is sufficient in itself to prove the evolution of that particular order of mammals, and a similar story could be told of other orders of that class and of many other groups, both high and low.

Since the time of Darwin and Wallace, the doctrine of evolution has permeated and revolutionized every department of human thought, and as a scheme of creation is immeasurably more reasonable than its crude predecessors. So far from belittling our conceptions of a Supreme Intelligence, it adds immensely to the dignity and wonder of the universe.

FREDERIE BREWSTER LOOMIS

AND NOW,

I dream that man will come to have an understanding of God. A knowledge that through God's Spirit all creation exists. Humankind are in this world of the universe as a part of this Great Spirit: God's Spirit is an Ether Spiritual presence which dwells in you as in all creation. Proof of a beginning is our existence: there is no end. Life shall be understood. As when, two thousand years ago, what you have learned to call God, spoke to you through Jesus, this same voice shall be heard again and again throughout Eternity.

BILL HUTH

MERRILY WE ROLL ALONG

The remains and the impressions of ancient animals and plants found in the rocks—the fossils discovered by our geologists—show a gradual multiplication of species from age to age, and a gradual increase in the complexity of forms and functions. The story told by the record of the rocks is a story of progress from the simple to the complex, from the lower to the higher. The earliest true vertebrates were the simplest fishes; next came the higher, more complex fishes; then the amphibians, able to live both in water and on land. The reptiles, such as lizards and crocodiles, came next; then birds, which, as the record shows, were derived from the reptiles; and the mammals, some of them, such as the apes and man, of very recent origin. The geologic record reveals strikingly the great story of evolution, and discovery after discovery is making the details of that record more clear and more convincing.

FRANCES MASON

THE TONGUE

A formal education has evaded me for many reasons. Probably the most significant could be my unending search for the absolute truth and an end or completion to understand what life is all about. There has always seemed to be something left of the secret of Being. Those who have spent their lives in search of this same truth or exposure come so close at times and I can feel in my heart that they are going to complete their explanation of what it is all about and then they, as the others leave you hanging. I fully understand that just as for me and you these great men of science and religion get closer to the answer than you or I, but the answer is always lacking. I feel that it will come. Just as the answers to life have slowly exposed themselves to us and this vast amount of knowledge we have accumulated is stored some where in our body. In order to keep this knowledge we have to continually search through it and add to it just as we have to do in our attic. If we don't check it out once in a while we will soon forget what we have learned so far.

I feel that our most significant organ next to our brain is our tongue. Besides the Parrot and a few other birds that have the ability to talk: humankind seem to have the only ability, as far as mammals go, to speak. And man seems to be the only one that understand, most of the time, what he or she is saying: A true message. This ability, I am sure, is the reason for the development of a part of our brain that none of the other creatures or parts of creation seem to have. By learning the art of speech I am sure that our brain had to develop in a different way in order to store the vast new amount of ideas and explanations told to us by others. Not one thing can compare with this except the brain Itself. Just as the body builder builds muscle the brain builder builds brain.

I feel quite sure that if the dog or apes or fish or birds had our ability, and which could be theirs with a tongue, which enabled them to talk, would soon develop what we as humans feel is ours exclusively. On this subject birds seem to be the nearest to accomplishing the art of speech: they sing their sentences. While in the confidence business in the south eastern part of the United States and Cuba, my wife Maxine had a canary. This bird was a real singer, maybe it was because she fed him good, maybe the area he lived in or it could have been living around us. We had pigs, at times, under the house, cats and even lizards would come in and catch spiders. We also had a mockingbird that would come around and sit outside in the Magnolias and copy singing what the canary sounded out. I also had a tape recorder that I taped the canary's songs on and also the copying mockingbird. One day while playing the mockingbird's mimics I noticed, as I watched the canary listening to the mockingbird's song on the recording, that he cocked his head at one particular sound the mocking- bird made. Everytime I played the tape with the same sound he re-acted in the same way. Although when the mockingbird showed up and I played the same tune to him he never re-acted in anyway. My thought on this was that the canary knew what he was talking-singing about and the mockingbird was only copying and had no regard or desire to understand what the song was all about. Kind of like me, I know all the melodies, but not one complete set of lyrics.

All of creation has its own language, even the trees speak through the breeze, true there's no, in our minds, relative conversation but it's as good as a growl or a meow or a whistle or a bark. Many mammals have learned words and sentences from us. Now since we have learned a method of living under water in the lakes and oceans, we have been able to talk with the mammals that live in the water and we have also achieved a manner in which to train them. I feel sure that what we have been able to do with them will soon show up in almost all of our fellow members of creation.

I'm, sure at the time of Arron and Moses that there were many different causes of "death" which were not understood at that time. One of the most profound could possibly have been what we now call strokes or cerebral hemorrhage. When these occur, the mouth takes on a grabbled sound of mixed or words which are undescribable, or lack any common sense. It's very indescribable what the witnesses of this sort of thing were going through as they watched and listened to the victim as she or he waggled their tongues making all sorts of sounds which were only heard from the mouth of someone who was going through this "death" causing convulsion. These type convulsions and this same type of garbled speech were common with this before "death" happenings. I can relate to these same circumstances as I have been witness to situation. Because of this type of speech before death, I feel sure that this was another of the hocus-pocus spiritual happenings that were born out of these "death dealing" circumstances. This could have been where "speaking in tongues" had its origination. What more could they have asked for; a person uttering his last words in tongues which only God could understand. I have also listened at places of worship where many have mimicked these same sounds in an effort to communicate with their different gods.

On realizing the extreme value of the tongue in making sounds which came out as speech and which were understood by most who were listening, the tongue took on a vital organ which had to be controlled in some way by those who were speaking, or by those who were listening. Many times in the ruff-and-tumble days of Moses a person could lose his life by stones being thrown at him just as tomatoes are used today. Thousands and thousands of times the tongue has been referred to as the double edged sword. "Motor-mouth" is a common expression by someone who thinks another talks too much.

The Tower of Babel was used by Moses and Arron as an excuse for the many different dialects which were spoken throughout the world then. Even these have caused the tongue's muscles to make it almost impossible for one nationality to properly pronounce certain words of the other's language. Talking in tongues was a proper speech for those who hung around Babel after the tower was destroyed by God. And babble, along with babel covers exactly what Moses and Arron wanted it to; to talk incoherently or foolishly or nonsensically, rant, rave, drivel, mouth, gush, maunder, run on, gossip, murmur, chat, chatter, prattle, prate, twaddle, gabble, tattle, jabber, palaver, blurt, patter, clack, beat the gums, run off at the mouth, talk off the top of your head, bat the breeze, rattle on, gab, cackle, blab, burble, sputter, blather, gibber, blabber, clatter almost everything, just don't talk things over that pertain to me, my religion, our God or my brother Arron.

BILL HUTH

THE EVOLUTION OF THE BIRD

ACCORDING to the story of evolution the various kinds of animals that we see to-day are not the descendants of like animals that were suddenly created in the forms they now have; some of them, at least, are the descendants of animals of far different structure and habits. We can observe that no animal is precisely like either of its parents in all respects, but in order to demonstrate certainly the larger changes covered by the theory of evolution we should have to watch carefully the natural breeding of some particular kind of animal for a long time—for thousands or even millions of years. We can get no such evidence as that, so we must turn to evidence of other kinds.

The alternative theory to that of evolution—the theory of special creation—assumes that each kind of animal was created in the form in which we now see it. If every kind of animal had this mode of origin we should expect to find that each one is a perfect machine, with all its parts arranged in the best possible way—that is, in the simplest and most effective way to perform their coöperative functions. But we find that all animals, regarded as pieces of machinery, are imperfect; each represents an attempt, more or less successful, to adapt a pre-existent structure to some new use. If the doctrine of evolution is true we should therefore be able to show that many of the present useless or anomalous struckept warm. The heart of a frog beats nearly twice as fast at 68° as at 50°. Thus if an animal is to be equally active under all climatic conditions it should be able to keep its body at a constant fairly high temperature.

An animal actually warms itself by moving its muscles, and it is able to keep its muscles going only by burning oxygen taken from the air, or, rather, by the natural process of breathing. The maintenance of a high bodily temperature uses up a good deal of food, and for mere economy it is desirable to reduce the quantity required by providing the animal with a coat that will allow its heat to escape very slowly. This is the original reason for the fur that covers a mammal and for the feathers that cover a bird. Probably one of the first steps in converting a reptile into a bird is to change its scales to feathers. Yet when we compare a wing or tail feather of a bird with a scale it seems at first impossible that the one should have come from the other; but the first feathers of the chick—those which it grows while it is still in the egg—consist of very short scale-like quills, whose ends fray out into fine plumes. These feathers are formed from the upper layers of the skin in exactly the same way as the scales of lizards are formed; indeed, they differ from such scales only in being longer. Between these incipient feathers and those which we know as quills we find all intermediate stages.

In order to enable the bird ancestor to utilize fully the increased activity made possible by its higher body temperature many changes of its structure were necessary. One of the most important of these has to do with the heart. A lizard can run very fast for a short distance, but it then collapses, completely exhausted, whereas a mammal or bird can hardly work so fast and so long that its muscles will no longer contract. This difference is due to the fact that the mechanism for sending a supply of oxygen to the muscles is much better in the bird or mammal than in the lizard.

The heart of a bird consists of two pumps, placed side by side. Into one of these pumps, that on the left side, blood full of oxygen comes from the lungs. This blood is then pumped forward through a great tube, which turns over to the right side of the animal and gives off blood vessels to all the muscles and all parts of the body except the lungs. All this blood, after being deprived of its oxygen, goes back to the right side of the heart and is then sent to the lungs to get a new supply of oxygen. The most peculiar part of the whole mechanism is that the great main vessel, the aorta, instead of lying in the middle line of the body, where we should naturally expect to find it, actually crosses from the left to the right side. We can explain this anomaly at once when we examine the structure of a crocodile. The heart of the crocodile is very like that of the bird, but in the crocodile there are two great vessels, one coming from the left side of the heart and the other from the right side. These cross one another in the middle line of the body and then pass upward until they join and form a single aorta. The vessel from the left side persists in birds, because it conveys only oxygenated blood, but that from the right side, although it is found in the very young chick, is blocked up in the full-grown bird, because it conveys impure blood. Thus the heart of the bird is better than that of the crocodile in that it supplies to every muscle the maximum quantity of oxygenated blood, and its peculiarities are explained by the structure of the heart of the crocodile.

By changes of the kind described, the bird ancestor was able to maintain prolonged and uniform activity. The first use to which it would put this power would naturally be in swift running in order to catch the small animals on which it lived and to escape from its enemies. It therefore began to run on its hind legs like a kangaroo. There are only two ways in which it could maintain such a bipedal gait—either by carrying its body upright or by having a long tail to balance the head and body. The bird ancestor adopted the latter plan. It had originally a spraddling walk, the feet being turned out and kept far apart. This gait made necessary a foot like that of a lizard, in which the big toe is the shortest and the fourth toe much the longest, so that the claws all lie on a straight line, at right angles to the direction of the movement of the body. The increase in the length of the toes is gained by an increase in the number of bones or joints that support them; the first has two joints, the next three, and so on, the fourth having five.

Any animal that must run fast must draw up its feet until they lie under the body, and at the same time the foot must be so shaped that the middle toe becomes the longest and the second and fourth, which lie on each side of it, become of equal length. This is the shape of the foot of a bird, though birds still retain five joints in their fourth toe, although they gain no advantage by so doing. These five phalanges are inexplicable if the bird was created as it stands, but they are easily understood if the bird was evolved from a reptile.

Any animal that uses its hind legs entirely for running or for such simple movements as scratching, that has its feet near the middle line of the body, and that runs fast, tends to simplify the structure of its foot by fusing together bones that were originally separate. In all ordinary animals the ankle joint is made up of many small bones and is, even in ourselves, a point of weakness. In a small chick these bones are represented by cartilage, but as the bird grows up they become separated into two groups, one of which fuses with the shin bone and the other becomes part of a single bone that corresponds with the bones of the arch of the foot and supports the three large toes. Thus the bird gradually comes to have a simple and very strong foot.

All animals that run very fast run on their toes, and the bird ancestor, in order to run more conveniently, permanently raised its heel from the ground. This change enabled the arch of its foot to be lengthened, so that with each stride it covered more ground.

Our bird ancestor has thus become a creature with long legs and a tail, capable of running very rapidly on its hind legs alone, the fore legs and hands being carried in the air. Such an animal, if it makes full use of its speed, must be able to use its hands for capturing prey and for carrying food. In order that it may do this its fore legs should remain rather short and the fingers should be provided with claws for holding the prey securely. But a clawed foot that is used for holding a struggling animal is an encumbrance if it retains all five fingers, and we find in many reptiles, even nascently in a crocodile, that the fourth and fifth fingers become smaller and smaller and finally disappear. A hand that is used for handling food must be capable of being turned about, and the arm that supports it must be freely movable. This stage of bird evolution is represented by many of the extinct reptiles that are called dinosaurs. These animals are not the direct ancestors of the birds, but they are close relatives, which went part of the way with them but, probably because they never developed feathers, were unable to make the last great step and begin to fly.

As the bird ancestors became able to run faster and faster they soon reached a stage when, like kangaroos, they travelled in a series of great leaps. They must then have extended the distance covered in these leaps by stretching out their freely movable fore legs so that they acted like the wings of an aeroplane and came to support more and more of the weight of the body. Their value as wings may have been greatly increased by the fact that they were already covered with feathers, which formed a fringe along the hinder edge of the arm, exactly as the scales do to the hind leg of a crocodile. The use of the fore legs as wings in this way, however, merely supports the front end of the body and head and leaves the tail trailing behind. But just as scales cover the whole of a lizard so feathers covered the whole of the bird ancestor, including its tail, and if these feathers retained the arrangement which they had as scales and followed the crocodile pattern they formed lateral keels along that organ. These keels will also act as wings, and if they become large enough are quite capable of supporting the weight of the hinder half of the animal.

Even after the bird ancestor had progressed to the point where it could rapidly move by a series of leaps, whose length was increased by gliding, while it gained speed by an increase in the power of the hind legs, the transition to true flight was a great step, which depended on the adoption of a flapping action of the wings. Such flapping, to be effective, must be regulated in a definite way, the course taken by the wing in its down stroke being different from that which it follows as it is brought up again to begin a new stroke. This perfect regulation of flight can be most easily assured by so shaping all the faces by which the bones move on one another and so arranging the ligaments by which they are tied together that no other movements are possible. In the modern birds this end is accomplished with great perfection by an arrangement that results in the fusion of all the originally separate bones of the palm of the hand and the wrist into a single element and a corresponding fusion of the two bones of the forearm.

But the adoption of a life in the air not only requires a modification of the physical structure of the wings and body but affects the relative importance and even the character of the senses as well as of the brain that makes use of the information they afford. The sense of smell becomes much less useful to a flying animal than it was to a crawling animal, which carries its head so close to the ground that it can recognise the presence of other animals by the odours which they leave behind them. We therefore find that the nose of a bird—its sense of smell—is in no way better than that of a crocodile.

In order to utilize fully the improved senses and to adjust them delicately to the conditions on which flight depends modifications must be made in the structure of the brain. The brain of a bird consists of the same parts as that of a crocodile and resembles it very closely in its fundamental arrangement, but that part (the cerebellum) which coördinates the muscular movements and adjusts them to the conditions under which the bird finds itself is much larger and is more complicated in structure. The part of the brain that is concerned with vision is larger, and the part that is concerned with smell is smaller.

More important in some respects is an enlargement of the front part of the brain to enable the bird to bring together there all the information that comes to it from its senses and to decide on its behaviour in the light of the memories of past events that are stored there. It is to the development of this part of their brains that birds owe all those competencies in building their nests and in caring for and protecting their young which have long endeared them to moralists.

In this account of the origin of the birds I have dealt only with certain selected parts of the body, and with only a few of the changes in the conditions of life. For example, I have not referred to the modifications of structure that are needed to enable a bird to perch on a twig. The features that I have mentioned were selected because they can be explained without presenting too much detail, and because they show the retention in birds of features whose presence is due to their existence as useful modifications of like features in their reptilian ancestors. The points considered may be summarized as follows:

1. The bird owes the presence of two, three, four, or five phalanges, or joints, in its toes to the fact that its reptilian ancestor, owing to its straddling gait, required toes that increased in length from the first to the fourth, whereas in the bird the second toe, which has three phalanges, and the fourth toe, which has five, are actually of the same length.

2. The bird owes the anomalous manner in which its aorta crosses from the left to the right side of the center of the body to the fact that in reptiles only this one of the pair of aortae transmits pure blood.

3. The bird owes its possession of only the first three fingers and not, as might have been expected, the fourth and fifth fingers, to the fact that the first three fingers are those which are most useful in the reptilian ancestor for clasping food between the two hands.

4. The bird owes the character of its brain to its descent from an animal having a brain like a crocodile.

This list might be extended indefinitely, but these four examples are sufficient to show that the peculiarities of the bird's structure—the points in which it seems to be clumsily constructed—are at once explained as relics derived from its reptilian ancestor.

If the explanation that I have given above is true we should be justified in believing that the oldest known bird, called *Archaeopteryx* (Fig. 1), which is known to us by two fossil skeletons, one in the British Museum, and the other in Berlin, is intermediate between the reptiles and the modern birds in its structure. *Archaeopteryx,* a queer lizard tailed bird, was found in rocks at Solenhofen, in Bavaria, which are in age nearer to the first rocks in which we find remains of ordinary birds than the rocks that were being laid down at the time when the change from reptile to bird began. We should therefore expect to find, and we do find, that it is nearer in structure to ordinary birds than to the reptiles. It has, for example, fully developed feathers, but it is in many ways a distinctly intermediate form. The hind foot,

FIG. 2.

"*Archaeopteryx* was considerably smaller than a crow, with a stout little head armed with sharp teeth (as scarce as hen's teeth was no joke in that distant period). While he fluttered through the air he trailed after him a tail longer than his body, beset with feathers on each side. Everyone knows that nowadays the feathers of a bird's tail are arranged like the sticks of a fan and that the tail opens and shuts like a fan. But *in Archaeopteryx the feathers were arranged in pairs, a feather on each side of every joint of the tail,* so that on a small scale the tail was something like that of a kite; and because of this long, lizard-like tail this bird and his immediate kith and kin are placed in a group Saururae, or lizard-tailed."—Lucas.

although it agrees in its proportions with the hind foot of other birds, has three separate bones in its arch, similar to the bones that are found in its reptilian ancestor. In the wing the bones of the forearm, the wrist, and the fingers are all separate, and the fingers end in big claws, so that they may have been used for capturing and handling food in exactly the way they were used by the reptilian ancestor, but in a way that no other bird does. Instead of the horny beak of a bird, *Archaeopteryx* has a row of little teeth that are exactly like those of a lizard. But one of the most interesting features of *Archaeopteryx* is its bony tail, which is longer than the rest of its body and along which there are two rows of quill feathers (Fig. 2). In the stage that is here represented, in which the wings were not big enough and not rightly placed to support the whole weight of the body and the tail feathers had to carry the hinder part of the body, the long tail was a necessity, but an unfortunate one, because it made it impossible for *Archaeopteryx* to fly with the perfection exhibited by such modern birds as the eagles and seagulls. An eagle rises from the ground by a few powerful strokes of its great wings and then, as soon as it has reached a certain height, it stretches its wings outward and upward, holding them motionless, except for tiny adjustments of their tips for steering, and soars away in gradually widening circles until it finally may become almost too small to be visible. Then, seeing a small animal on the ground, it partly closes its wings, falls headlong to the ground, stops suddenly by expanding its wings and short tail, and lands directly on its prey. No aeroplane can copy this dive, sudden stop, and accurate landing, because an aeroplane becomes uncontrollable when its speed falls below very high speed; and when it stops it must run for some way along the ground.

During the last few years we have learned of the condi-

FIG. 1.—The earliest known bird, *Archæopteryx macrura*, Upper Jurassic, Solenhofen, Bavaria. (Restoration by Heilmann.)

"The earliest bird known, although obviously a bird, with wings and feathers, differs in many ways from modern birds. It has large teeth; its tail was not a fan, but a double row of feathers on either side of a long pointed axis of a bone; and most remarkable of all, it had on its wings, besides feathers, three separately movable fingers ending in claws, by whose aid it doubtless scrambled about through the branches. In all these ways it was less fully adapted to aerial life than are modern birds."—Julian S. Huxley.

tions under which soaring flight is possible, and those aviators who have flown engineless aeroplanes (gliders) for hours at a time have been copying the eagles and giving us valuable information about the difficulties of the process, the chief of which is perhaps the recognition of the upward-directed currents of air on which it depends.

The other difficulty, the inability of the aeroplane to fly very fast or very slowly, depends on the long tail that all present-day aeroplanes have. As soon as it becomes possible to control an aeroplane that has no tail, or a very short one, man may be able to copy the manoeuvers of the eagles. From this point of view *Archaeopteryx* corresponds to a present-day aeroplane, the modern birds to the aeroplane of the future. It is certain that *Archaeopteryx* was clumsy, incapable of hovering over one spot and of alighting on a definite perch.

Archaeopteryx was therefore far inferior to the modern birds (see fig. 3) in its power of flight. It was clumsy, ill constructed, and lacked that perfection of form and motion which makes the sea gull a constant source of delight. Is it credible that a bird that was miraculously created in a moment should be so imperfect? Is not the imperfection of its machinery an evidence of evolution? Is it not more reasonable to recognize in *Archaeopteryx* a necessary stage in the long process by which a crawling reptile was gradually converted into the perfect flying bird of to-day?

I have here tried to bring together facts about the birds that bear witness to their evolution from more primitive ancestors. Comparative anatomy, embryology, and palaeontology unite in telling the same story. They agree in testifying that the bird is to-day a highly specialized descendant of some reptilian ancestor. Is it at all probable that there has been collusion among these witnesses and that their testimony is false? Must we not admit that the scientific research of to-day, no matter how much its results have disturbed mediaeval prejudices, has led us and is still leading us to a more reasonable conception of the order of Nature and of the true mode of creation?

DAVID MEREDITH SEARES WATSON

WHAT'S NEXT?

"The whole architecture of a bird skeleton, indeed the whole internal anatomy, is unquestionably a modification of a primitive reptilian type." —William King Gregory.

EVOLUTION "ISM"

"If we admit that species have changed, and are changing at the present time, that is all the principle of evolution implies. The evolutionist stands for and believes in a changing world, and unless you, the reader, believe in a fixed, unchanging world, you, too, are an evolutionist. Evolution is merely the philosophy of change as opposed to the philosophy of fixity and unchangeability."—H. H. Newman.

DOUBLE WORDS FOR THE DOUBLE CROSS

Let's start off with Adam, or Atom, or Atum, how about Aton or even Adhan; pretty bad if your hearing wasn't just right. Then there's babel and babble, or relegion or religon. These are the real winners - sole and soul; I've had it out with them. But for now relegion and religion are my concern. First let's take the word, and it is a word according to the dictionaries, re: again, anew, rebuild, renew or reborn. Now let's look a ligion: nothing there, no meaning, no word. How about legion: a very large number: a multitude. So if you say religion, it means nothing because there is no word for ligion and re by itself means rebuild or put back together. But now take relegion. We have a meaning for re, and we have a meaning for legion; a large number, a multitude, or to gather. Also it means the principal unit of the Roman army. I'm sure the Jewish people didn't want anything to do with this meaning. So if there is no word ligion, to put re in front of it means nothing still. But if you put re in front of legion it means to re-gather, or re-assemble which fits in more with what we conclude that ligion is; a word with no meaning at all.

BILL HUTH

CONNECTING AND MISSING LINKS IN THE ASCENT TO MAN

Two truths impress themselves strongly upon the mind of the student of animals, first, the continuity of life, and second, the immensity of time during which that continuity has endured. Our conviction of the continuity of life is justified by an overwhelming host of facts, obtained in part from observation of existing creatures, in part from the study of the geologic record. The evidence of the continuous succession of living forms is conclusive to modern scientists, having in mind, as they do, the classic experiments of Spallanzani, Redi, Tyndal, and Pasteur, which proved, apparently beyond the possibility of dispute, that all life is derived from preëxisting life. The evidence afforded by the fossil record, on the other hand, does not at first sight seem so convincing, for the absence of certain "missing links" in the chain of life is striking. Were the tale a short one, the apparent gaps in the record would be of greater relative moment, but, in view of the great length of the revealed story they become comparatively insignificant.

These gaps are in reality comparable to missing pages in an ancient and partly mutilated volume, which, although their absence surely mars the perfection of the whole, it nevertheless cannot destroy our understanding of the story that is told if we possess the imagination of the historian or the prehistorian. When this imagination restores the links in the chain of life we call them hypothetical; the links thus restored are not random guesses, but carefully predicted stages, and many such predictions have been shown to be approximately correct by fortunate subsequent discoveries. It is therefore not too much to expect that more such predictions may yet be verified through the intensive systematic exploration of to-day, which is very different from the often unorganized search of the past. Indeed, the memorable finding of the planet Neptune, after the astronomer's mathematical calculation had shown that it must be in a certain part of the heavens at a certain time has had its parallel in more than one recorded verification of a like prediction in palaeontology. To trace complete continuity in the succession of living forms that have inhabited the earth would necessitate the finding of a representative of each generation, a thing that is manifestly impossible. There must always be breaks in the record, but we are disposed to insist that the breaks should not be very great, and that they should not occur at highly critical stages in evolutionary advance. To find serene security in our faith in the continuity of life we must still await discoveries that will bridge certain breaks.

The origin of living matter—of organic matter—from the lifeless material of the inorganic world was a most momentous step, for it led ultimately to the peopling of the globe with its countless hosts of animals and plants. When, where, and how life began, however, we do not know, although much purely academic discussion has been waged about the question. As students of the origin of the earth must assume the preëxistence of matter and energy, so students of organic evolution must assume the existence of something organic to evolve, but science is silent on the great problems of first causation.

CONNECTING AND MISSING LINKS

Our fossil record begins with the dawning of the Palaeozoic era, some 500,000,000 years ago, but the date of the origin of life was vastly more remote, for when the fossil record actually begins not only is life fully manifest, but the numerous animal stocks that constitute its invertebrate division are already well established. These earliest known animals are far more diverse and complex than the most primitive imaginable organisms, some of which yet exist—such as the slime molds and the bacteria described as prototrophic (literally "first feeding"). Thus the missing links in the early evolution of animal life constitute for millions of years the entire chain. There is, however, no doubt in the scientific mind that the oldest known fossils imply, with the assurance of certainty, a long antecedent evolution, much of which, in spite of the fact that palaeontology is silent, can be deduced from the sister sciences of comparative anatomy and embryology. The reason the record anterior to this time is blank is because of the nature of the organisms themselves. Composed as they were largely of soft parts with no limy skeletons, or shells, they apparently left little to be preserved in the rocks of ancient time. Consequently, save for certain remarkable impressions, yet to be described, little of the actual nature of these ancestral types is known, except by inference, until they had established what has been called the lime-secreting habit, which was formed by animals in mid-Cambrian time, by plants somewhat earlier. True, there are masses of limestone, iron ore, and graphite in the pre-Palaeozoic rocks, all of which are regarded as largely of organic origin, but, although their presence is indirect evidence of the existence of organisms in this remote time, it reveals nothing of their nature.

One notable exception lies on the flank of Mount Wapiti, near Field, in British Columbia, where, in a small area of slate of Lower Cambrian age, the most marvellous impressions of delicate and fragile organisms have been found—mere films of carbon against the slate—with all their detail of structure preserved with the utmost fidelity. These impressions show many of the relatively high invertebrate types, such as worms, crustaceans, and echinoderms, and here there might well be preserved the ancestral stock out of which the backboned creatures arose, although, so far as actual discovery goes, that record is much later in time.

Except for a persistent type of being variously known as *Amphioxus* or *Branchiostoma,* which to this day inhabits the shallow waters bordering the continents, there is again no trace of the important link connecting the vertebrates with their invertebrate forebears. To those of us who are prehistorically interested this is perhaps the most coveted of all the missing links, for its discovery will help to settle arguments that have arisen in favor of this or that method of origin, about which so little is really known.

In Silurian time there came the sharks, known largely from their fossil teeth, some so much like the teeth of modern persistent survivors that the external form and habits of life of some of the early species may be safely surmised by analogy, although others are unlike any that now exist. Of their gradual evolution, especially in the favored lines which were to produce the higher fishes, sufficient, perhaps, is known. Out of one armored group, however, there were to arise the land-living vertebrates. Here again there is more hypothesis than observed fact, and especially desirable would be the discovery of fishes whose fins show potentialities, either in their structure or their implied use, or both, of giving rise to the shore-adapted foot, the point of departure of all terrestrial progression. There is at Yale a single footprint, which speaks volumes to those who read its lesson aright, for it is the most ancient record of a terrestrial vertebrate known to science, although there is still some difference of opinion as to its correct interpretation. But it is baffling in its obscurity, as it tells little of the structure of the impressed foot and nothing at all of any other part of the owner's frame, except, again, by inference. This footprint (*Thinopus*) is a most important link in a weak part of our chain; but other links, which thus far are known but vaguely or not at all, are necessary for the chain's integrity.

The old-time emergents—those that first left the water—were lung-breathers, of course, but they still went back to the water to bring forth their young. Living amphibians, such as frogs and salamanders, yet do this each recurring season, and gill-arches preserved in certain fossils, along with shore-adapted hands and feet, show conclusively that such was their ancient custom.

The passage into the next higher group, that of the reptiles, implies the loss of gill-breathing in the young and the consequent laying of eggs ashore. This passage has given rise to a large, complex egg, which can both nourish the developing embryo and allow it to breathe as well, until at the time of hatching it emerges as a miniature snake or turtle or crocodile, according to its kind, but never as a form reminiscent of the ancestral fish in shape or habit. Here the transitional types are surely known; the uninitiated can not tell whether they are amphibian or reptile, the difference lying in certain technical details of structure that are discernible only to the expert; and here, therefore, our chain has all the requisite strength of continuity.

Links of great interest are those that connect reptile and bird on the one hand and reptile and mammal on the other. Huxley, years ago, spoke of the birds as "glorified reptiles," and his descriptive term is still very apt, for the birds are merely reptiles that were especially adapted to aerial navigation. Nature essayed this adaptation twice with the reptiles, one product being the pterodactyls or winged reptiles of the Mesozoic era. The pterodactyls and the birds have much in common, even to certain minute details of pneumatic bones with comparable openings of communication one with another: lightness of skull, with a precocious consolidation of cranial bones, and loss of teeth and their replacement by a beak in later forms. The marked distinctions lie in the mechanism of flight—by wing feathers in the birds and by a bat-like wing membrane in the pterodactyls. The two groups are doubtless derived from the same or a related non-flying ancestry, and much of their similarity is probably due to community of habit and its reaction on the mechanical structure.

The links leading to the pterodactyls are still unknown, for our first fossil record of these creatures are the remains of forms that had already attained sustained flight; of the first steps in that direction we have no direct knowledge. Not all conditions of life, however, are equally susceptible of preservation in the fossil record, and this is particularly true of flying forms, for a prerequisite to fossilization is complete burial either in water-borne or air-borne sediments, such burial as befalls flying forms only in very exceptional circumstances. A single locality, a quarry near Solenhofen, Bavaria, worked commercially for limestone used in the art of lithography, has given us nearly all the flying creatures of Middle Jurassic age that we know. Here are pterodactyls, some of which have the delicate wing membrane preserved in perfect detail; but these are already highly specialized fliers. Here also are the first known birds, feathers and all, and among the whole list of connecting links known to us none is perhaps more satisfying. Three individuals are represented, one by a single feather and the other two by complete skeletons, except that one is headless. This headless form, now in the British Museum of Natural History, is the famed *Archaeopteryx;* the other, in Berlin, shows sufficient distinction in its preserved parts to warrant a new generic name, *Archaeornis.* Both are classed as birds largely because of the feathers and implied powers of flight; in other respect they are reptiles. *Archaeornis,* in fact, has been called a reptile in the disguise of a bird, and the skull might well belong to a reptile, never to a modern bird. Here a none too efficient flight is already attained, but here again the history is missing, and our ideas concerning the origin of bird flight must still remain hypothetical. The scientific visualization of this pro-avian is of a light-built dinosaur-like form, running freely on the hind limbs and occasionally taking to trees for soaring leaps, sustained largely by partly modified reptilian scales. Out of such a scale in turn would evolve that masterpiece of nature, the feather—a prophecy that may in time be realized by fortunate discovery in the older rocks. From *Archaeopteryx* to the toothed birds of the Cretaceous period is again a considerable unbridged gap, during which the bird became essentially modernized except for the retention of teeth in the jaws, which may well have been due, however, to a habit of eating fish. We can well imagine intervening stages, showing gradual adaptation to greater efficiency in flying. The shortening of the lizard-like tail of *Archaeornis,* the feathers of which are arranged in a row on either side, to the fan-like tail of the pigeon or crow produced a vastly better device for manoeuvering in flight. The hand has consolidated the old free grasping fingers into a better and stronger wing, and the skull has changed its character in many details. A succession of drawings comparable to a moving picture film might well be made to show these changes with a fair assurance that, one after another, forms showing the comparable actual stages might be found. Here the lack of the connecting links does not seriously disturb the evolutionist, although their discovery would be an event of profound interest.

Our own lineage lies of course in the mammalian line; hence the dawning of mammalian life is of intense personal concern. Here we know much of the truth, for many of the stages have been revealed. The chief distinctions that separate the mammals from their reptilian prototypes are the peculiar methods of nourishing the young both before and after birth, much of the internal mechanism of the mother being directly or indirectly modified as a result of this habit. That the early mammals were egg-layers is attested by the retention of the egg-lying habit in the monotremes, such as the duckbill of Australia, the lowliest of existing mammalian forms, which, with many another evolutionary laggard, is a veritable living fossil, existing in a place remote from the busy competition that impels advance. The palaeontologist cannot trace the development of these diagnostic mammalian characteristics, for they are limited largely to soft parts. The preserved strictly mammalian features that may be compared with features possessed by members of the ancestral reptilian group are few. We find differentiated teeth, which are embedded in the rear of the jaw by more than one root; a single bone in the lower jaw, which is directly articulated with the temporal bone of the skull; and other minor details, largely changes toward greater simplicity.

Back at the beginning of the age of reptiles there existed, mainly in the Southern Hemisphere, a group of reptiles known, from their differentiated dog-like teeth, as cynodonts, from κυνός, a Greek word for dog. That they were not mammals is shown by their complex jaw, which consisted of several bones on either side, and by their retention of a quadrate bone, as it is called, between the jaw and skull. Both of these features had been possessed by reptilian, amphibian, and fish ancestors as far back as we can trace the bony skeleton, and are thus a firmly established heritage of the race. Why the simplifying? And where shall we find the missing bones? These questions have given rise to much argument. At all events, the changes occurred concomitantly with the assumption of other diagnostic features, and in a comparatively short time. The Triassic rocks yield jaws of creatures so near the dividing line between reptile and mammal that only the most intensive modern research has decided their status as reptiles, although for years they were considered primitive mammalian forms. Mammal and bird each have warm blood, which means not only a heat-controlling mechanism, but clothing (hair or feathers) as well. Feathers are seen in *Archaeornis,* but fossil hair is still unknown in association with ancient mammals, so that in them warm bloodedness cannot be proved, though it may be assumed from analogy. From the mammal-like reptile sprang the reptile-like mammal, out of which true mammals in turn arose. Much of this history is recorded in hundreds of tiny jaws and teeth recovered from rocks of the reptilian age. Certain links in the chain are missing, but the main evolutionary lines are indicated by tangible evidence, not alone by inference.

Mammalian divergence has followed several lines, all traceable to a few parent stocks. These lines led to the hoofed cohort, to the clawed carnivores, and to those groups which went down to the sea and became adapted marvellously to life in the great waters—whales and sea cows of diverse lineage. Other feebler folk became the inhabitants of trees, and these belonged to a very ancient group, retaining certain primitive traits of limbs and teeth which left them far behind the others in evolutionary status. But out of these, largely through mental supremacy, were to come our forebears and, later, actual humanity itself.

Certain recorded links couple lines which to-day are so divergent that students of recent animals would hardly imagine their relationship. Compare the manatee, of spindle-like body, broad swimming tail, flipper-like fore limbs, and no visible hind limbs at all, with the majestic elephant, conspicuous for his peculiar features of limbs, head, and trunk. What community of structure do they show? And yet there has come out of the Libyan desert of Egypt a fossil form that is neither one nor the other but partakes somewhat of the nature of both—a synthetic swamp-dwelling type, whose descendants came to the parting of the ways and went severally to the sea and to the land—to grotesque aquatic efficiency on the one hand and to noble terrestrial majesty on the other. Here the link has been found. Were it lost, who could tell the tale of this strange kinship?

One of the most completely known of all evolutionary chains is that of the horse-like forms, though here again the hypothetical five-toed ancestor is still undiscovered. We could, however, restore him in detail, so complete is our evidence of the course of equine evolution, and, if the chain includes other missing links we can formulate their characteristics with mathematical certainty, as well as their distribution in time and place. Our material for tracing the horses is so abundant that it resolves itself not into a single chain, but into many, all of which have arisen from the original stock to follow lines that may be closely parallel, or, again, widely divergent, but invariably in response to climatic and vegetative environment. Most of these chains have come to a natural end at one time or another, so that we now have but two, or at most three, terminal lines, represented by the true horses, the asses, and the zebras. The lineage of each of these forms is apparently traceable back through millions of years, until the lines ultimately merge.

But it is with the lineage of Man that we are chiefly concerned. Out of some unknown line of primitive mammals arose those small beasts of prey whose lack of prowess necessitated their feeding on such feeble folk as they could overcome—worms, slugs, insects, and small lizards. These beasts are represented in the living fauna by the Insectivora, such as the shrew, the mole, and the hedgehog, which represent ancient lines of descent and display rather conservative forms. As primitive offshoots of this group came on the one hand the stronger feral animals, known as the Carnivora, and a group of tree folk, the Primates. Where the original home of the Primates was we do not know, but we strongly suspect that it was some circumpolar region of salubrious climate where extensive persistently green forests afforded them asylum and an abundance of easily obtainable food. Fossil plants of Eocene age comparable with those living in Cuba today have been found in Greenland and Spitzbergen. It is safe to assume, therefore, that the climate of those lands was then also comparable. From early Eocene time we have a definite fossil record of Primates in England and in Wyoming, some of the American forms being well-nigh identical with the living lemurs or half-apes of Madagascar, an identity implying that these again are persistent types, or "living fossils." In North America these early Primates lingered throughout the Eocene but disappeared there when a change of climate destroyed the tropical forests that had been their home. They migrated through Central America to South America, where their somewhat altered descendants yet persist. In Europe they had a similar history except that they reappear from some Asiatic or African fastness, flourish for a while, and die out again until the coming of Man. The monkeys of Gibraltar are really African, despite their foothold in this outlying rock.

Living monkeys, both tailed and man-like, are found to-day in both Africa and Asia, and although all are derived from the same root as mankind, the great or man-like apes, especially the gorilla and chimpanzee, are our next of kin, being separable from man, aside from spiritual values, by what in other creatures would be considered inconsequential details. It is only our lack of perspective that makes us imagine a considerable gulf between us.

Human and simian relationships are not yet clearly traceable; they are inferred from close structural and functional similarities, the distinctions being referable to differences of habit and habitat. That man and the great apes are cousinly descendants from a common stock all scientists believe; they do not believe, as some imagine, that man is an exalted ape or that the ape is a degenerate man, though what one should call the common progenitor, if not an ape-like form, I do not know. This precursor of ours has been long sought, and he has been called *the* "missing link," as though one link could constitute a chain covering many thousands of years of descent. *Pithecanthropus,* the form found by Dr. Dubois in Trinil, Java, in 1891-92, stands not far from the ancestral form and more than any known today deserves this title. This type, which existed half a million years ago, was then already erect, though not perhaps of god-like carriage, showing how remote was the attainment of this human characteristic; but in form and capacity of skull he lies about midway between the gorilla and the Neanderthal type of man that lived in Europe some 25,000 to 40,000 years ago. In that sense he is a connection, but he is not a common ancestor. Heidelberg man and his lineal successor, the man of Neanderthal, constitute the first and last links of another chain whose slow change is unrecorded for three hundred and fifty thousand years. But these forms are evidently not in the line that led to modern man, although the Neanderthal type was yet alive when Man appeared on the European stage, 25,000 years ago. What lies back of our own species is still unrevealed, but we are probably out of an unknown though extremely ancient Asiatic stock. That Asia is the birthplace of humanity most authorities now agree, except the few who, because of the primitive character of the African natives and the antiquity that was Egypt, infer an African origin for the higher races. This inference seems unwarranted, for, although the negroes are primitive in most respects, in others they show a higher specialization than either the yellow or the white races. Fossil forms are now coming to light, especially in the Siwalik region of India, which could be ancestral to the great apes, and possibly to man, but one looks for the final solution of this problem farther to the north, in the comparatively unknown Asiatic plateau.

That there are gaps in our revealed record of the continuity of life—gaps due in part to incomplete exploration, in part to natural causes—is manifest; but that the record is sufficient to uphold the principle of continuity is equally manifest. Sedimentary rocks form the repository of this record, and sedimentation is always locally discontinuous because of the wearing down of the earth's surface until the force and carrying power of the streams have well nigh ceased. This wearing down of the land is followed periodically by its re-elevation through crustal movement, with consequent rejuvenation of the streams, which begin once more the endless round of scouring, carrying, and depositing. When the crustal movement is so profound as to be dignified by the name revolution there may be marked organic change, stimulated usually by stress of climate, which, either directly or indirectly, has a vast influence upon life. These times of change are therefore critical, both in the elimination of old types and in the acceleration of the evolution of persistent types, so that the whole aspect of nature as revealed by the fossils is profoundly altered. It is not at all surprising, therefore, that the older naturalists, whose orthodoxy caused them to adhere to the doctrine of special creation, imagined a succession of great catastrophes, by which the older faunas were completely destroyed and life was recreated twenty-seven times, the number chosen corresponding to the greater divisions of geologic time. This same explanation was applicable locally to the lesser breaks in the record. But geologists are now convinced of the uniformity of physical conditions throughout geologic time and of the all-sufficiency of the observable phenomena of the present world—of changes in temperature, of rain, snow, and ice, of erosion and geochemical action, of earthquakes and volcanic activity—to account for any and all of the changes, however great their apparent magnitude, in the geologic past. And as a necessary corollary of this doctrine of uniformitarianism in the physical world comes that of continuity with evolutionary change in the organic world. That the evidence for this organic continuity seems meagre is due in part to our lack of perspective, in part to our prepossession with false conceptions or pseudo-conceptions, and in part to our proneness to magnify imperfections that merely mar but do not destroy a most magnificent, clearly unified, and deeply impressive moving spectacle of creation, which at length makes Man the heir of all the ages.

RICHARD SWAN LULL

SOON OUR MIND WILL ACCEPT

"The testimony yielded in the superficial layers of the earth and in the caves, embracing hundreds of specimens of the fossilized remains of man, more or less ancient, more or less complete, invariably, without a single exception, testify to the gradual *ascent of man* from a lower to a higher state, gradually dropping one primitive bit of anatomy after another until the high, intelligent, fully human aspect is attained."—H. F. Osborn.

NEVER AGAIN

"Determinate evolution is victorious all along the line. It only remains for us either to accept the inevitable, or to take up a position as belated outsiders, living in dreamland away from the practical realities of science, rethinking the childish thoughts of primitive folk."—C. Lloyd Morgan.

TRUTH - IT'S OURS TO UNDERSTAND

"The time has come when scientific truth must cease to be the property of the few—when it must be woven into the common life of the world." —Agassiz.

NOW IT'S TRUE!

For seventy years, since the age of three, I have searched for the answer to life and Its heaven. At the age of four my life became half empty with the loss of my father. I was told at the time I would be able to join him in heaven. I'm almost positive I have thought over the question of, does heaven really exist? as much as or maybe more than many people. Will I be able to get in? Will I know him when I see him? Will he know me? I'm a lot bigger now. I wonder if my friends will be there, and they say dogs are not allowed. It's gonna seem funny now, I'm over seventy and he was forty-five when he supposedly went to heaven.

These questions could go on and on and if the truth were known, they are all wasted thoughts. Just a few weeks ago the pope for the catholic church informed his followers that "the theory" of evolution is no longer a theory but a true fact. Facts, as we know them, are just what we feel at this time to be a true answer to a particular question or supposition.

I believe my first suspicion of heaven came to me when I was a boy and they told me my little dog could not go to heaven. As a matter of "fact" they told me that there were no pets in heaven. Now this made me have to choose between my pets and my dad. I couldn't seem to bring myself anymore to a real meaning for heaven. In truth I started up-rooting all of my feelings towards heaven and started seeking my own solution to the questions of life and the here-after. It was never my luck to see any phantasm or angels, I couldn't even dream about them no matter how I tried.

Around the age of ten I began hearing stories about Charles Darwin and evolution. At this age I was spending most of my time with my pet mice, owls, pigeons, dogs and every other wild thing imaginable. I never, at this age, realized what Darwin and his "story" were really all about but it sounded a little more like heaven than what I had been offered.

Life went on and I continued to learn more and more about this "theory" that I was told was nothing but a bunch of scientists and their lies. I had to listen to both sides however because I was very positive of one thing: I was going to KNOW what all of this was about. I kept ripping at heaven and trying to find another answer to what seemed to be getting to be a much larger question the older I grew. The only Bible study I had done was in Sunday School and most of that was Noah's Ark and the other fables of Jewish history.

Finally around the age of thirty I decided to read the Bible. It took me about three months of every day reading. The first thing that really perked me up was The book of Ecclesiastes, 3: 18-22. Solomon said he felt God would make man understand that man was a beast and that the dog had a heaven too. The words were put together in a little different way but they meant what I said. Then I went on and Isaiah said, in his own way that the new moon and the sabeth and sacrificing were all something that his God hated. As I read on I began to see that some of the people who wrote in the bible were only story tellers, and most of them made up. About half of the real writers were really against the churches of their time and wanted reform. But the real purpose of the whole thing seemed to tell me to FEAR their God.

NOW IT'S TRUE

So I became more and more involved with evolution. I read Darwin. I didn't understand much about it but like the bible I tried and tried and spent more time in the dictionary than I did reading the books.

After 33 years of from heaven to hell I printed my first real thoughts about religion and evolution. My publication, "The Voice of Creation", hit the streets. I didn't do too bad, I received a letter from the University of Iowa asking me to send them copies and information about the "Voice". Which I did. So my life continued on its way. I made money for a while and then I studied for a while and then I hustled for a while and then read more and sometimes I wondered if it would really ever be possible to show others the fallacy of Adam and Eve and the truth of evolution; both physically and mentally. I spent another forty years and at last the pope had to cop-out. The facts of evolution are just too much to handle and from that day on the children will no longer hear about the garden of Eden. Now I am sure it won't be much longer for many of the "stories" of the story tellers to be handled just as their story of how creation got started, and tossed out just as it has been. I sincerely believe that Moses and Arron were marching to a different drummer when he told them theirs were the chosen people. If they were chosen, their God didn't have much to offer besides to be scared as hell of him or he would "grab the earth and turn it upside down and shake all of the chosen ones off into the wild blue universe." At least he kept his word; after the flood he put his rainbow in the sky as a deposit or covenant that he would never flood the earth again. I can only say, I hope the pope knows what he is really saying. I'd hate to have to go out in the wilderness for another seventy years.

BILL HUTH

THE LINEAGE OF MAN

Early Stages of Life

THE story of the evolution of man may at some distant date come to an end, but apparently it never had a definite beginning. In order to get a reasonable historical perspective let us open the story at a period which, for the sake of illustration, may be thought of as a billion years ago. At that time by far the greatest advance toward the human type had already been made, for living matter, according to our present evidence, was already in existence. All the myriads of years before that in which the components of living matter had gradually been built up had passed. The birth-throes of the central sun, which under the gravitational attraction of another passing sun had whirled out great tidal arms (somewhat like the streamers of a pin-wheel) and given rise to the planets, had long since been forgotten. The earth and the other planets had settled down nearly into their present orbits, and the surface temperature of the earth was not materially, if at all, higher than it is to-day. Moreover, the waters of the earth's surface had long since been gathered together into oceans, the continents (whatever their outlines) were already in balance with the oceans, and the well-nigh eternal round of rock erosion, deposition, consolidation, sinking, and uplifting had gone on for hundreds of millions of years. Occasionally there were terrific disturbances of the earth's crust, the older sedimentary rocks were soaked and honeycombed with the molten rock coming from below and were squeezed, mashed, folded, contorted, baked, and partly melted, so that if they ever contained any traces of life, such traces would have been hopelessly obliterated, the only possible signs of organic life being the beds of graphite occasionally found in the oldest known rock formations.

Such is the picture suggested to the geologist by the study of the oldest known rocks, now exposed again in eastern Canada, the Lake Superior region, and the Adirondacks as a result of hundreds of millions of years of erosion, but in former ages lying as the "basement complex," beneath tens of thousands of feet of later rocks.

Passing over various doubtful traces of living organisms in the oldest sedimentary rocks (such as the famous *"Eozoön canadense,"* which may be a mineral formation), we come to certain markings occurring in the Proterozoic limestone formations, which were determined by the great palaeontologist C. D. Walcott as the calcareous secretions of algae. In the upper levels of the "Belt series" of formations in Montana, exposed on the side of a mountain and lying nearly eight thousand feet below the Cambrian, or lower part of the Palaeozoic, Walcott found the fossilized traces of worm burrows and trails, seemingly of segmented annelid worms.

Hence even far below the bottom of the Palaeozoic system, which to the earlier geologists marked the utmost lower limits of the record of fossil life, we come upon animals of marvelous complexity compared to their one-celled starting point. But where the fossil record fails, the "chain of beings" still in existence apparently preserves some of the main steps in the elaboration of higher living types out of single living cells. For even to-day there are forms of life that afford strong evidence for the following outline.

The single living cells from which higher life started were for a long time independent creatures, capable of assimilation, growth, and subdivision. After a time, when the daughter cells adhered together, a more or less spherical colonial type appeared, finally attaining regional subdivision and division of labor. Then one side of the sphere grew faster and a pushed-in ball appeared. Hereafter the inner layer served for the digestion of the food and became the primitive gut, while the outer layer not only held the bag together but developed sensory and contractile powers, as in the lower jellyfishes. Meanwhile the puckered skin around the mouth grew out into feelers and stinging tentacles. All this looks simple, but the organization of each individual cell was an affair of unimaginable complexity.

Certain jellyfishes began to give up their free-swimming habits and to squirm or crawl on the muddy bottom. Presently the diffuse "nerve net" throughout the body began to be drawn together into definite tracts, the squirming movements finally became more prominent in one direction, and locomotion in a head-and-tail direction was already begun.

The Origin of the Vertebrates

The vertebrate animals originated millions of years later, and there is as yet no general agreement as to what group of invertebrates gave rise to the vertebrates. Professor E. B. Wilson teaches that the vertebrates (or chordates) belong to that great branch of the animal kingdom in which the mesoderm, or middle layer of the three primary cell-layers, arises from outpockets from the primitive gut, as it does also in the echinoderms (the starfish group) and that all the articulated animals, such as arthropods (crustaceans, insects, and arachnids), annelids (worms), mollusks and other groups belong to a series in which the mesoderm buds off from a single cell or pole cell. Professor William Patten, on the contrary, holds that this distinction is not a fundamental one and that the vertebrates have been derived from some early member of the arthropod series, such as the fossil eurypterids.

Both the vertebrates and the arthropods are many-jointed animals provided with an elaborate locomotor apparatus, and both have a highly complex head, which has apparently developed through the growing together of a number of originally independent segments. According to what may be called the orthodox view, all the resemblances between vertebrates and arthropods have been independently acquired in these two great groups, for both had to solve many similar mechanical problems in the perception, pursuit, ingestion, and digestion of their food. According to Professor Patten, on the other hand, the arthropod mechanisms were attained first and afterward were changed to form the vertebrate ground-plan of organization along lines which he has inferred, but which the orthodox reject as requiring too many hypothetical stages between arthropods and the oldest vertebrates known.

The Earliest Chordates

Recently Professor Johan Kiaer has described many beautifully preserved fossil fish-like forms from the Silurian rocks of Norway. These fossils belong to a group of animals, hitherto known chiefly from the Silurian and Devonian rocks of Scotland and Russia, which are commonly called ostracoderms. Some (including *Cephalaspis*) were flat-bodied like skates; others were shaped more like ordinary fishes. The modern lampreys appear to be degenerate descendants of this group, which is also remotely related to the sharks and higher fishes. Professor Stensiö has collected very perfectly preserved fossil specimens of *Cephalaspis* from the Devonian rocks of Spitzbergen. Serial sections of these specimens have been studied by Professor Patten, who states that the radiating bony channels for the cranial nerves and many other architectural features of the anatomy of the head conform to the general plan seen in the head of the fossil eurypterids and other arthropods. Patten therefore argues that this new material has proved his theory that the vertebrates have been derived from the arthropod stock.

Whether the vertebrates came from very early arthropods or whether they were derived from unknown cigar-shaped forms that preceded both the ostracoderms and the existing lancelets (*Amphioxus*), it is at least certain that the earliest known ostracoderms already foreshadowed the higher vertebrates, including man, in the ground plan of their organization. Already they had the main chordate characters that are displayed in the human embryonic and foetal stages but that are masked in the adult human stage, namely, a notochord or elastic axis, above which is the central nervous system and below which is the primitive gut and heart.

Like all primitive chordates the ostracoderms swam head on, by throwing the long body into waves proceeding from in front backward. This undulating motion is produced by the rhythmic contractions of a series of zigzagging muscle plates ranged along each side of the body from behind the head to the base of the tail. Each zigzag is separated from the next by a partition of connective tissue which runs inward toward the notochord.

The ultimate unit of locomotion is not the zigzag muscle segment but the short red muscle fibre. Thousands of these little fibres are placed along the zigzag path of the muscle segment, each fibre being attached at its front and rear ends to the connective tissue partitions between the segments.

Each little red muscle fibre is a tiny "gas engine," consuming the oxygen of the blood stream; it is touched off, so to speak, by the nerve current which is conveyed through a nerve fibril from the larger nerves that pass down from the spinal nerve cord to the muscle.

All the complicated locomotor apparatus of the vertebrates, including man, has evolved according to clearly discernible stages out of this relatively simple ground plan. In support of this statement Nature supplies us with hundreds of different variations of this simple theme. The ultimate causes of evolution may be as mysterious as you like, the origin of the vertebrates from invertebrates may be obscure if you will, but the main stages in the evolution of the locomotor apparatus, from fish to man, are now a matter of record, and the same is true as to the evolution of the human skull, brain, and spinal cord.

The following story of evolution has not been built up like a system of metaphysics or philosophy out of abstruse untested reasoning; it is the plain result of many more or less independent lines of research and discovery pursued by geologists, palaeontologists, zoologists, embryologists, and other scientists for more than a century. Naturally, within the limits of the space available in this book I cannot review the evidences that have led to this general picture of vertebrate evolution.

Early Evolution of the Fishes

Whatever may have been the origin of vertebrates, by the time the Devonian period was reached a very great advance had been made toward the higher forms, for at that very distant time, probably half a billion years ago, there were already in existence shark-like fishes that resembled man in possessing the following important structural characters:

They were already true vertebrates, with the same main divisions of the whole body as in man, the same ground plan of the brain and spinal cord, the same type of segmentation of the vertebral column, the same general type of complex skull (to be described later); moreover, these ancient Devonian fishes and their modern representatives agree with early stages of the human foetus in the general plan of the jaws and gill arches and in the basic features of the digestive, circulatory, respiratory and reproductive systems. Hence the humble dogfish or shark, which is a relatively little modified survivor of the early vertebrates, is universally recognized by biologists as affording a true ground plan of human anatomy and physiology.

We do not know why the earliest fishes gave rise on the one hand to a line of forms that progressed to ever higher types and on the other hand to far more conservative ones, including the existing sharks and ganoid fishes, which represent, on the whole, an arrested or retarded evolution; but the fossil record of the vertebrates shows that in every geological age one finds the more conservative, less modified descendants of older stocks living contemporaneously with the far more progressive relatives. The science of comparative anatomy, in collaboration with allied branches of science, is able to decipher the evolutionary history of man from the vast mass of Nature's documents precisely because the more conservative forms of each geological age enable us to visualize the anatomical characters of older periods and to reconstruct, with progressive approach to complete accuracy, the steps by which the older conditions have given rise to the newer ones.

By lower Devonian time the vertebrate stock had already split up into the following main groups: first, the ostracoderms, an excessively ancient group, of which nearly all the known lines were then near the end of their career; second, the shark group, at that time relatively small, which was to become highly diversified in the next period and then to be crowded almost to the wall by the higher fishes; third, the true ganoids (actinopts), ancestors of the sturgeons, spoonbills, bony gars, and eventually of the higher or teleost fishes; fourth, the dipnoans or lung-fishes, and fifth, the lobe-finned or crossopterygian ganoids, to be described presently.

Among which, if any, of these groups are we to seek for the ancestors of the higher vertebrates, including man? The known ostracoderms of Devonian time were already specialized side branches, all far too late in time to be the actual ancestors of the higher vertebrates. The Devonian sharks were already giving rise either to the specialized side branches or to the ancestors of the modern sharks; the actinopt ganoids were already on the line of advance to the higher fishes; the dipnoans were already highly specialized in their dental apparatus and skull characters and well along on the line to their modern descendants. Only certain of the lobe-finned fishes seem to have the right combination of characters to be even nearly related to the direct ancestors of the land-living vertebrates.

What, then, are the broad characteristics of these interesting fossil fishes, some of which may lie relatively near to the line of our own ancestry? Some of them were long-bodied, pike-like fishes, with great, strong jaws armed with sharp teeth. Others were stout and heavy-bodied, somewhat like a sea-bass, with shorter jaws. One very specialized group that lasted for many millions of years was short-bodied, propelled by fan-shaped paddles and by a very broad tail. Of all these, only the first lot, including *Osteolepis, Megalichthys, Eusthenopteron* and their allies, appear to be near the line of ascent to the land-living vertebrates. In these strong-jawed, pike-like fishes each of the stout paired fins (corresponding to our arms and legs) was supported by an internal skeleton consisting of bony rods converging toward a single bone, corresponding respectively to our single upper arm bone (the humerus) or to our thigh bone (the femur). The fore paddles were supported by a complex shoulder girdle, parts of which correspond to our collar and shoulder bones; the hind paddles were supported on a bony plate corresponding to the lower bars of our pelvis.

The skull of these lobe-finned ganoids, like the human skull, was a complex of two very distinct sets of elements. The inner skull, or braincase, consisted of the bony trough surrounding the brain and of the bony shells or capsules surrounding the organs of smell, sight, and balance. The outer skull consisted of a shell of bones, derived, like the scales on the body, from the skin. In the earlier crossopts (such as *Osteolepis*) the outermost layer of the bony skull and scales consisted of a hard, shiny, porcelain-like substance called ganoin, but in many of the later crossopts this outer layer was lost, leaving a sculptured bony surface.

Among living fishes only the famous *Polypterus,* the bichir of the Nile, and a nearly related genus have any claim to be considered the modified descendants of these lobe-finned or crossopt fishes of the Devonian period. These interesting relics still retain vestiges of former air-breathing arrangements in their lungs or swim-bladders, but they now rely chiefly on their gills for aeration.

Origin of the Amphibia

The connecting links between the lobe-finned fishes and the amphibians of the coal forests are not yet discovered, but all the known earlier land-living forms agree in so many points of structure with the lobe-finned fishes that there can be little doubt as to their relatively close relationship. The alternative possibilities, either that fishes were derived from four-footed amphibious animals, much as whales have been derived from mammals, or that air-breathing fishes and four-footed animals represent entirely distinct groups of vertebrates, have been carefully considered by those best qualified to weigh the evidence and have been rejected for excellent reasons.

We must infer, then, that some adventurous pioneers among the early air-breathing fishes managed to crawl out of the muddy waters at times when either the supply of oxygen in the water or the supply of living food there was insufficient; that at such times these creatures wriggled along on their bellies much as some eel-like fishes do now under similar circumstances, except that they used the stout fore and hind paddles to increase their hold on the mud and to assist in pushing the body forward. From that time on, the stout fan-shaped fore-paddles began to be bent at the future elbows and wrists, while the hind paddles were bent in the opposite directions at the future knees and ankles; meanwhile the fan-shaped bony rods of the paddles broke into segments and gave rise to the bones of the fingers and toes.

In many of the earliest land-living vertebrates there were five principal rods or digits on each hand and foot, and possibly a small nodule or reduced ray in front of the thumb or the great toe, and another behind the little finger or the little toe; but these extra digits in most lines of animals were early reduced to vestiges, so that the five-rayed hand and foot became the standard form. Thus man in common with other vertebrates has inherited the basic pattern of his five-fingered hand and five-toed foot from the earliest land-living vertebrates, perhaps of the days of the Devonian coal forests. Man also owes to these Devonian fishes his very ability to breathe.

Man has, moreover, inherited from these amazingly ancient animals each and every one of his twenty-eight skull bones, as well as every other bone of his entire skeleton. But the earlier types possessed many *more* bones in the skull than man does. The late Professor Williston, of the University of Chicago, showed that as we pass from earlier to later vertebrates reduction in the number of the skull elements is the rule, although there is here and there an exception due to the fragmentation of one of the remaining elements.

Fishes, in common with all other vertebrates, have, on each side of the head, the three semicircular canals which act like spirit levels and assist the animal in maintaining its equilibrium. Below the semicircular canals is a sack supplied with nerves. A part of this sack corresponds to the true organ of hearing in man. The cavity of the human ear (behind the drum membrane) is represented in fishes by the cavity of the gill chamber. Fishes have no drum membrane, which first appears in the amphibians, but its place is occupied in the fishes by the bony shell or operculum covering the outer side of the gill chamber. When the air-breathing fishes were changed into amphibians the opercular series disappeared, leaving only a notch at the outer upper corner of the skull, upon which was stretched the drum membrane of the ear.

Thus, by the time amphibians originated from some air-breathing lobe-finned fishes, all the fundamental problems in the production of the ground plan of the human organization had been solved, and it is literally true that the oldest amphibians were on the whole much nearer to man than they were to the oldest known ostracoderms. Nevertheless, a host of improvements in the entire mechanism were still to be worked out, and to these striking advances we now turn as we pass from the oldest amphibians to the oldest reptiles; thence to the theromorph series of reptiles, which became more and more mammal-like; thence to the oldest mammals, onward to the most primitive placental mammals of the closing age of the dinosaurs; thence to the tree-shrews, lemuroids, monkeys and apes that grow steadily more like man; finally to man himself in the last few million years of the billion-year history of life.

Origin of the Reptiles

In their individual development or embryology the amphibians, like the fishes, went through a water-living stage of development, in which they had functional gills; but the earliest reptiles succeeded in laying their eggs and rearing their young wholly upon land and in this way were able not only to invade the drier uplands and many parts of the earth where water was scarce, but to avoid the intensive competitive warfare for living food that must have raged in the swamps of the coal forests.

The earliest reptiles were still so much like their contemporary relatives, the amphibians, in most of their skeletal characters that some of them were on or relatively near the borderland between the two classes. Such a form is *Seymouria,* from the Permian of Texas, a fine skeleton of which is mounted in the University of Chicago. The pattern of its skull bones, as seen from above and from the side, conforms closely to the primitive amphibian type and is an almost ideal archetype of every later skull, including that of mammals and of man himself. The same is true of the underside of the skull, including the arrangement of the numerous elements of the upper jaws, palate, and base of the cranium. It may be said of *Seymouria,* as of many other generalized forms, that, if we had not discovered them, we could have predicted their existence, so closely do they conform to the ideal conditions inferred to be the starting point for later advances.

The Mammal-Like Reptiles

We come next to the theromorphs, mammal-like reptiles of Permian and Triassic time, nearly all of which were fierce carnivorous animals. The most primitive of these were lizard-like forms from the Permian rocks of Texas and Russia; but some of them, called pelycosaurs, were specialized side lines. The most noteworthy advance by the earlier members of this group was the development of an opening, or temporal fossa, in the bony mask covering the back part of the skull on each side behind the eyes. The most primitive amphibians and reptiles had this temporal region covered with a continuous shell of bone, but in the mammal-like reptiles the tissue surrounding the jaw muscles (which lay just beneath this bony shell) seems to have acquired the power first of fastening itself to the surface of the bone and then of sinking into the middle of the bony area. Meanwhile the bone itself, while giving way in the middle of the area, strengthened itself around the margins of the area to which the jaw muscle was attached. In this way bony arches grew up around the margins, and an opening appeared toward the middle. Such openings have been developed in different parts of the temporal region in different groups of reptiles; some reptiles had on each side two temporal openings or fossae, one above the other, and other reptiles had none. The mammal-like reptiles had but one, which was surrounded in the later members of the series by the postorbital, parietal, squamosal, and jugal bones. This temporal fossa may be traced through the ascending series of mammal-like reptiles into the early mammals, thence into the tree-shrews, lemurs, monkeys, apes, to man himself. The bony bar below this opening, variously called the jugal or malar or zygomatic bone, may similarly be traced upward to man.

We likewise owe to these lower predatory reptiles the reduction in number of the wedge-like pieces that finally grew together to make a single vertebra or functional unit of the backbone, there being four pairs in the early amphibians but only two fully developed pairs (neurocentra and pleurocentra) in the higher mammal-like reptiles and mammals. These aggressive and progressive animals also took many other steps toward the mammalian type of skeleton, including the human type, most of them correlated with their improved running powers; for whereas the earlier forms had crawled almost on their bellies with sprawling arms and legs, the cynodonts, or later mammal-like reptiles, carried their bodies well off the ground, almost like the more primitive mammals, as we know from the detailed form of their limb bones.

Among the improvements introduced by the cynodonts was the reduction of the phalangeal formula (that is, the number of bony segments or phalanges in the different fingers and toes of each forefoot) to the mammalian number, as indicated in the following table:

NUMBER OF PHALANGES IN EACH DIGIT

Digit	I	II	III	IV	V
In earlier reptiles	2	3	4	5	3
In cynodonts	2	3	3	3	3
In primitive mammals	2	3	3	3	3
In primates (including man)	2	3	3	3	3

Similarly in the hind foot the primitive reptilian phalangeal formula was: $\frac{\text{I} \quad \text{II} \quad \text{III} \quad \text{IV} \quad \text{V}}{2 \quad 3 \quad 4 \quad 5 \quad 4}$; but the formula in the higher mammal-like reptiles and mammals was: $\frac{\text{I} \quad \text{II} \quad \text{III} \quad \text{IV} \quad \text{V}}{2 \quad 3 \quad 3 \quad 3 \quad 3}$

Certain of the lower mammal-like reptiles show transitional stages, for in them one phalanx of the third digit and two phalanges of the fourth digit were much reduced in length and appear to be almost on the point of disappearing (Broom). Very possibly the reduction in the number of phalanges in cynodonts was associated with the newer method of walking more on the ends of the fingers and toes.

Thus it may be seen that man has inherited not only the number of his fingers and toes but even the number of the small bones on each finger and toe from the mammal-like reptiles of the Triassic period.

Man also owes to the higher mammal-like reptiles a whole series of structural improvements in his skull, teeth, and jaws, which the limits of space here compel us to deal with in a summary fashion, for the teeth of these reptiles are already reduced to two sets, corresponding to the milk teeth and the permanent teeth of man, and are differentiated into incisor, canine, premolar, and molar teeth, as in the mammals, including man. Moreover, the cynodonts clearly foreshadow the mammals in the progressive predominance of the dentary or tooth-bearing bone of the lower jaw, one on each side, which in the mammals is the only surviving one of the numerous separate pieces found in the lower jaw of reptiles.

Origin of the Egg-laying Mammals

The higher mammal-like reptiles take us almost to the threshhold of the mammals, and they almost exactly divide the structural differences between mammals and primitive reptiles. Meanwhile the typical or modernized reptiles, including the turtles, lizards, crocodiles, and dinosaurs, acquired divergently specialized characters that carried them far away from the mammal-like reptiles, which, as we have seen, were progressing toward the mammals. Most of the mammal-like reptiles certainly became extinct, but we must suppose that certain others went on and gave rise to the mammals through such intermediate links as *Dromotherium* and *Microconodon,* tiny animals from the Triassic of North Carolina, each of which is as yet known only from one-half of its lower jaw. At any rate, the mammal-like reptiles as a group realize all the predicted characters for the ancestors of the mammals, and those ancestors, when they are more fully known, should be intermediate between some of the mammal-like reptiles on the one hand and the mammals on the other.

The mammals apparently originated during late Permian or early Triassic time, in a semi-arid region that was subject to extreme changes of temperature. In South Africa, one of the homes of the mammal-like reptiles, the polished surfaces of the older Permian rocks clearly indicate the presence of great continental glaciers, which are generally accompanied by periods of extreme cold and aridity, followed by warm interglacial periods.

The chief difference beteen a typical mammal and a typical reptile is that the mammal has far more perfect devices for regulating its own body temperature and thus compensating for changes of temperature in the environment. As mammals have an active diaphragm and similar improvements, they can generate more heat in proportion to their weight than reptiles, and in their hair they have a superior substance for retaining the body's heat; also, by means of sweat glands, they can lower their own temperature through evaporation.

Typical mammals have also become able to hatch their eggs within the body, eliminating the egg-shell and bringing forth their young alive, but most reptiles conserve the old habit of egg-laying, which is also retained to-day by the most archaic living mammals, the duckbill (*Platypus*) and the spiny anteater, both of Australia. These two ancient forms connect the mammals with the extinct mammal-like reptiles in several respects. The opossum, kangaroo, and other marsupials, taken as a whole, represent an early stage in the higher method of reproduction, for their young are born in a very immature state and are subsequently developed in the mother's pouch.

Origin of the Placental Mammals

In all the higher mammals, such as the dog, cat, horse, cow, elephant, monkey, ape, and man, the internal development is carried further than in the marsupials, and a more extensive connection between the growing embryo and the womb is established by means of the after-birth or placenta. Here is another fundamental structure which man obviously owes to his mammalian predecessors.

The same is true of the breasts of the female. From the presence of rudimentary breasts in male mammals (including man) people sometimes infer that the remote ancestral forms must have been hermaphrodites, but all available evidence indicates rather that in the ancestral mammals the sexes were just as distinct as they are to-day.

The presence of breasts in the female of the human species and their ability to secrete milk for the nourishment of the young were among the facts which led the great Swedish naturalist Linnaeus, in 1758, to list mankind within the class first called by him Mammalia, and the presence of only a single pair of breasts, together with other considerations, led Linnaeus to group man as a member of his order of Primates, which included also the apes, monkeys, lemurs, and bats.

To return to the fossil mammals: Our knowledge of the very long period after the mammals first appeared, during the ages when the dinosaurs and other reptiles dominated the scene, is extremely meagre. From a study of recent mammals Huxley predicted that the remote common ancestors of all the highly diversified placental mammals would be found to be small insectivorous forms, not unlike some of the recent insectivores, such as the Madagascar tenrec (*Centetes*) in general appearance and habits. From a study of the teeth of later mammals Professor Henry Fairfield Osborn likewise predicted that the ancestral placentals, living perhaps in the Lower Cretaceous period, would have teeth of the generalized insectivorous type. Quite recently Dr. Roy C. Andrews and his colleagues have contributed important evidence in favor of this view by discovering in a Lower Cretaceous formation in the Gobi desert in Mongolia the fossil skulls of several kinds of small mammals. Some of these combine features of the later insectivores and primitive carnivores and thus appear to afford a generalized pattern for the divergent evolution of the insectivores, carnivores, herbivores, tree-shrews, and primates, all of which are first definitely known to have lived in Palaeocene time in North America, at the beginning of the Tertiary period, or so-called Age of Mammals.

Palaeontologists are confident that these already diversified mammals were not suddenly created in Palaeocene time, holding that they were derived by evolutionary changes from the more primitive mammals of the Cretaceous and Jurassic periods, from some of which they inherited certain striking characters of their dentition.

Once we have passed out of the obscurity of the very long period during which the reptilian hosts dominate the fossil record and the mammals remain far too inconspicuous, we come to the much fuller record supplied by the fossils of the Age of Mammals, estimated by Barrell to be about sixty million years in duration, with its six great epochs—Palaeocene, Eocene, Oligocene, Miocene, Pliocene, Pleistocene.

Origin and Evolution of the Primates, Including Man

Throughout this enormously long period, which was short, however, compared to some of its predecessors, the fossil records are relatively abundant for some great orders of mammals, such as the hoofed mammals, and extremely meagre for the primates. In western North America at the beginning of the Palaeocene epoch, some sixty million years ago, there lived relatives of the existing tree shrews, and in the next higher beds (Lower Eocene) we find the ancient relatives of the lemurs and tarsioids, which are found also in the Eocene of Europe. In the Lower Oligocene beds of Egypt have been found two lower jaws of extraordinary interest, one (*Parapithecus*) combining the characters of the tarsioids and the anthropoids, the other representing a primitive pro-anthropoid ancestral to the gibbons and perhaps to the branch leading to the higher apes and man. In the Miocene and Pliocene beds of India and Europe we find the broken jaws of possibly a dozen kinds of anthropoid apes, some of which (*Dryopithecus*) appear from the details of their teeth to be closely related both to the existing anthropoids and to man. In the Upper Pliocene beds we find possible traces of early man in the shape of crude flint implements; in the Pleistocene beds have been found the remains of many individuals of the Neanderthal race in Europe; also the famous skull of *Pithecanthropus,* in Java. In the closing stages of the Pleistocene epoch the modernized *Homo sapiens* appears.

Although the fossil record of the evolution of the Primates

is meagre it tends to show that the various groups appeared in the following succession: (1) tree shrews, (2) lemuroids and tarsioids, (3) monkeys, (4) pro-anthropoids, (5) diversified anthropoids, (6) primitive man, (7) modernized man. But this is also precisely what one would predict from a comprehensive comparative study of the surviving families of Primates, with special reference to the structure of their brains, skull, teeth, hands and feet, and other parts of the body. The main branches of the order are all represented to-day by surviving members. By making comparisons first within each group and then between groups, including both the fossil and the recent forms, it has been possible to decipher the main record of evolution of the brain, teeth, and various parts of the skeleton.

Taking the series as a whole, it shows a remarkable gradation of forms and structures, culminating in various side branches and also in man. It will remain for future palaeontologists to correct errors and to amplify the details of the process of evolution, but the general sequence of events has been worked out independently by a number of investigators, whose results yield a remarkably concordant, consistent story. Thus Keith has shown that when we pass from the monkeys to the gibbons, which stand near the base of the anthropoid-man radiation, we find that the gibbon has already effected profound readjustments of the viscera and skeleton to its habit of sitting upright and of brachiating, or extending the arms upward and leaping from branch to branch. Keith finds that on the whole the gibbon is nearer to man in this internal readjustment to the upright position than it is to the lower primates. Elliott Smith and his students in England and Tilney in America have worked out the sequence in brain structure from tree-shrew to lemur to monkey, gibbon, orang, chimpanzee, gorilla, man, and they find progressive changes culminating in the enormous expansion of the brain and intelligence in man. The present author has published a series of works on the recent and fossil Primates, dealing especially with classification and phylogeny as founded on studies of skull structure, dentition, and various parts of the skeleton, in which the same general sequence as that derived from a study of the brain has been worked out. With the collaboration of Dr. Milo Hellman, the writer finds that, as formerly suggested by Dubois, the Miocene and Pliocene group of anthropoids called *Dryopithecus* clearly foreshadow both the modern anthropoids and man, and that certain species of this broadly inclusive "genus" appear to be at least not distantly related to the actual common ancestor of man and chimpanzee. Remane, after the most profound studies, has shown that the canine teeth and anterior premolars of man retain many tell-tale evidences of derivation from a stage in which these teeth were like those of certain female chimpanzees; and the great anthropologist Schwalbe left as his legacy to the world a masterly analysis in which he endorsed the conclusion that man and the chimpanzee are the offshoots of an ancient common stock. Professor Henry Fairfield Osborn, although formerly doubting the derivation of man from an arboreal stock, now finally accepts the remote arboreal origin of man and the derivation of both man and anthropoids from a common anthropoidean stock. He therefore differs from the present writer chiefly in inferring that even as far back as Lower Oligocene time the cleavage between man and apes had already begun. But whether this cleavage took place in Lower Oligocene time or somewhat later, the conclusion remains, based upon a vast accumulation of evidence, that the higher anthropoids, especially the chimpanzee and the gorilla, are man's nearest surviving relatives, and that the remote "common ancestor" of perhaps ten million years ago was a tailless, partly tree-living, pro-anthropoid, in many respects far more like a young female chimpanzee than like a modern white man.

Conclusion

The natural egotism of man made him easily credulous of the story that the first man, although made from the dust of the ground, was also created perfect in the image of God. The knowledge that man has struggled upward to his present estate from less intelligent animals is still practically denied to the majority of mankind.

The gospel of evolution as outlined above is not the writer's invention; it has not been built up, like early systems of religion, in an endeavor to propitiate the gods without; it is simply a very condensed outline of what Nature is gradually revealing to those who carefully examine her records. When man fully realizes what he has come from and the long, slow steps by which he has reached his present condition, he will be better able to apply intelligent measures toward correcting his infirmities and toward guiding his evolution along profitable paths in the future. One can do no better than quote the noble words of Charles Darwin:

> We must, however, acknowledge, as it seems to me, that man, with all his noble qualities, with sympathy which feels for the most debased; with benevolence which extends not only to other men but to the humblest living creature; with his god-like intellect, which has penetrated into the movements and constitution of the solar system—with all these exalted powers—man still bears in his bodily frame the indelible stamp of his lowly origin.

WILLIAM KING GREGORY

LET'S GET BACK TO THE SOUL

Not seeming to be important because it has already been agreed on that the Soul is the thing that belongs to the individual whose body it represents, however it's how we get into heaven and still have the features by which others can recognize us. It's our outside appearance and inside it's that someone else we talk with and make most of our very important decisions. Just plain spirits aren't worth too much because we can't see them, but our Soul is something else! After all, what is the Spirit worth to us if we don't have what we were raised with, our appearance? Getting down to the basics, our body is sort of hard to be recognized unless it's some sort of freak thing ten foot tall or one foot long legs and six foot long arms with no body; just anything like that anyway? I'm sure that if the majority of the people had a more accepted type body most of them would kind of like to have the same design; maybe a little different but still almost standard. Even with all, or almost all of the bodies being almost the same, we still have a big problem. Color! Color seems to be one of our most serious concerns. In the beginning, each color seemed to think their color and personal "self" was the correct color to be; others were inferior. There are four of these colors that seem to present the biggest difficulty, and many times cause bad problems: red, black, yellow and white. All four of these seem to stand for separate races, yet they have all learned to speak one language, dress almost the same, eat the same basic foods, live in houses which are almost the same, drive cars and use the basic tools which are all the same. Many of these basic colors have crossed over into the other colors and have began a sort of melting pot; otherwise slowly blending together. I'm sure that as evolution continues on the path it has already chosen and as the crossing over continues, color will cease being any sort of an issue at all. What a relief: How long this will take is any "bodies" guess, but we've been going through this evolutionary process for a couple of billion years so far, so a few hundred or thousand more shouldn't bother us much now: so when this color problem is finally done away with we will only have our faces left as a means of recognition. Now we have to develop a way in which we will be able to keep these faces for ever and ever and into the here-after.

So here we are. Our color has been blended into one, our bodies are almost identical, but we still seem to be retaining our self-styled faces. Now it seems that we will be facing another major dilemma. What do we do about aging? Or showing the characteristics of increasing age; or in other words show signs of ripening. This process, which belongs to all creation, begins as something to be proud of. Our features are developed in a way that make us improve and show off our personalitites. To a certain age, which varies with each individual, we become very attached to our facial designs and then old age becomes a problem. Some go and have a redesign, or "facelift" which is a temporary improvement, but not really as much as the individual anticipated. So many of these lifts and a few times at the dye shops and then most people who have tried to save their youthfullness wish they had never started it in the first place. So here we go; it's time for heaven at last. We're sinless just like they say we've got to be to get through the gates and everyone is waiting for us to "show up": arthritic joints, dyed hair, liver spots and wrinkled skin. What more could you not want? So whoever is in charge takes all of this away and makes you all brand new and ready for the streets of gold. Nothing else to worry about; a new body, new bones and a new face that looks just like the old one and now we're set for eternal life, where everbody lives forever. There's no way you can lose your life again up there; this is for keeps: FOREVER! Never an end, timeless time. You also have a few options to pick from. You can select the age you want to be, and stay at that age forever, you can select an aging range, too, if you would like you can live a few years and then bounce back a few years, just repeating this same cycle over and over or every once in a while you can go back clear to the beginning. One very important thing though; you must always retain the same body and face. Remember this is going to be always. If you want you can select a stage of tranquility, which also lasts forever and you won't even know what's happening: this is really sort of like what you used to think dying was like, but a little different, you're still alive.

At any rate you still have your body, face and soul. You're in heaven and there's not a thing to worry about. Nothing to look forward too. Nothing to worry about. And it'll be like this forever!

BILL HUTH

EVOLUTIONIST?

People say you don't have a God. I say, look about you and your surroundings. He says; so! I say, today I can yell and scream about religion. I can print about religion. I can do everything I can to convince others that the mentality and physical shape of humankind has been brought about by what we now call evolution. It is this evolution that presented to me the God I now believe in.

In 1494 William Tyndale was born on the English, Continent. William became a biblical translator, Humanist and Protestant martyr. Prevented by ecclesiastical authorities in England from publishing an English version of the Bible, he began printing the New Testament in Germany at Cologne (1525) and completed it at Worms. He also worked on the Old Testament (1530-36). His translation later became the basis of the King James Version of the Bible. Seized by the authorities at Antwerp, he was condemned for heresy and burned at the Stake. Imagine trying to burn someone at the stake today! This is what the Evolutionist's God gave to His followers; FREEDOM. Tyndale was only translating the Bible; not burning it!

BILL HUTH

THE HUMAN SIDE OF APES

For some peculiar reason the animal kingdom includes several kinds of creatures that are remarkably like us in a great many ways. Everyone has noticed the amusing resemblances between apes and men, but few are aware of the numerous and close similarities between them that are revealed by a thorough comparative study. Bone for bone, muscle for muscle, nerve for nerve, we are remarkably close counterparts of our anthropoid relatives. Even in the structure of the brain, which is, perhaps, our most distinctive anatomical peculiarity, there is, as Prof. G. Elliot Smith has remarked, no essential difference, except in degree of development, between ape and man. To be sure, we have a much larger brain, and the so-called association areas are more extensively developed, but in brain structure we differ less from the higher apes than these differ from the lower members of the monkey tribe.

Now mental development and brain development are closely tied together. We stand far above the apes in the development of our minds, and no one is wise enough to gauge the degree of our mental superiority from a study of the structure of the brain. A comparative study of brains, however, would lead us to infer that the ape stands nearer than any other animal to man in mental endowment. And this inference is abundantly justified. Nevertheless, the gap between ape and man is wide. Mr. Darwin, with his well-known candor in giving the fullest weight to objections against his theory, described it as "enormous," yet in his *Descent of Man* he endeavors to show that the differences in mentality between man and the higher mammals are not fundamental. Like the mental differences between human beings, which are also enormous, they are differences of degree and not of kind. Consequently we may readily conceive that the human mind may have arisen by an orderly process of development from a mind of lower order.

We are not obliged, however, to regard the evolutionary origin of the human mind as merely a plausible possibility. The evidence for its evolution is much the same as the evidence for the evolution of the body. Resemblance in fundamental features of structure and in method of individual development is justifiably regarded as a strong indication of a common descent. When we compare man with a chimpanzee and note the same form of the external ear, the same rudimentary ear muscles, the same slope of the hairs on the arms and legs, and countless other similarities, even in little useless features of structure, we find the only reasonable explanation of these similarities in the conclusion that they are inherited from some common ancestor.

The same argument applies to the mind, though minds are not so well suited for detailed comparison as bodies. We think that we know something about the human mind, but most of our knowledge lacks the accuracy and precision of the subject matter of our big books on anatomy and physiology. Our knowledge of the ape mind is much less complete. Until recently no one had ever made a really systematic study of the intelligence of the higher apes. Mr. Darwin did his best to bring together the available information on the subject when he wrote his *Descent of Man*. He had to content himself with the scattered and often rather casual observations of many naturalists. He collected several anecdotes that show the intelligence of apes, their power of imitation, their strong parental affection, their mutual sympathy, their grief over the loss of their young, and their services to one another in times of danger or distress. He showed that the apes share with us all the common basic emotions and that they express their emotions by much the same bodily signs. No one can read Darwin's book on *The Expression of the Emotions in Man and Animals* without experiencing a sense of the fundamental kinship of the human and the animal mind and of the likeness in their expression of the emotions. Anger, fear, affection, astonishment, grief, pride, disappointment, and disgust are expressed in much the same way, not only by all the varied races of mankind but by the apes and monkeys. The language of the emotions is a universal mode of communication. The frown has the same meaning in man and apes and is caused by the contraction of the same muscles. The broad similarity in emotional life and in its expression that we share with our simian relatives is as strong an evidence of common origin as the similarity in the form of the skeleton or of the brain.

We now know considerably more of the mental life of apes than was known in the time of Darwin. We are getting better acquainted with our simian cousins, and our more intimate acquaintance has led us to a more generous appreciation of their mental qualities. Many people habitually think of animals as prompted only by feelings that correspond to the lower passions of our own nature. The terms bestiality, animality, and brutality are terms of reproach. The words "ape" and "tiger" are synonyms of ruthless ferocity, the antithesis of everything we regard as worthy in human character. Many are doubtless prejudiced against the doctrine of evolution because they feel that the foundations of morality would be undermined if the conviction becomes general that human beings were derived from animal ancestors. They do not sufficiently realize that the good as well as the evil qualities and impulses of human nature have their counterparts in the animal world. Both men and animals are occupied chiefly in the work of maintaining and perpetuating life. This work involves, in animals and man alike, a due adjustment of efforts to promote individual welfare and the welfare of others of the same species. Most animals pay little heed to the needs of creatures outside their own family or social group. Human beings do likewise. We think little of exterminating animals to satisfy our own needs, or even for mere sport; but we picture the gorilla as a horrible and dangerous creature if he can be provoked into making an attack upon a human being. But why should a man be anything more to a gorilla than a gorilla is to a man? To his own associates this commonly misrepresented animal is a kindly creature having a creditable endowment of domestic and social virtues. So is the man-eating tiger and the prowling wolf. Toward her little group of playful cubs the lioness is an indulgent and self-sacrificing parent, ready to incur any danger to protect her own kind. From her viewpoint man is just so much potential meat for the support of herself and the offspring of her body. The lioness is a beast of prey and a natural enemy of the human race because the evolutionary process, or the Lord, or perhaps both, made her in that particular fashion. Man in turn regards the lioness as a dangerous creature—a creature to be ruthlessly exterminated to insure his own safety.

After all, man's superiority to the lower animals is due

primarily to his much finer intellect and to the greater variety and delicacy of his emotions and sentiments. Probably no animal can appreciate the beauty of a sunset or the charm of an attractive landscape. If we can rightly boast of any moral superiority over our less favored animal associates it is not because we are more devoted to our own kind or follow more faithfully the standards of our own particular group; it is because we are consciously moral and are able to make moral judgments and talk about right and wrong; and we can do these things simply because we have much better minds and a richer emotional life than our animal progenitors.

The fear that the foundations of morality would be undermined if it were proved that we are derived from an animal ancestry is eminently absurd. The foundations of moral life lie deeply rooted in the domestic and social instincts, which form the mainsprings of action in animals and men alike. We do not speak of sympathy, mutual helpfulness, or parental love (Fig. 1) as parts of our so-called "animal nature," although in consistency we should do so, for these traits are as much a part of the nature of animals as ferocity or greed. It is traits such as affection, sympathy, and group loyalty that constitute the basis of our moral impulses and sentiments. Our social and altruistic impulses are no less worthy of esteem if they are shared by less highly developed creatures than ourselves. Like the lower animals, we are in general sympathetic and helpful to our own kind. To our enemies and the enemies of our country we are hostile, and often cruel. All this is human nature. It is also animal nature. In man and animals love and antipathy, courage and cowardice, self-sacrifice and selfishness, loyalty and deception, play much the same part in determining behavior. We play the game of life less simply and crudely than the animals, but our fundamental interests in life are much the same.

In order to gain a clearer appreciation of this general fact let us consider briefly some of the mental and emotional characteristics of our next of kin in the animal world. As to keenness and vision, acuteness of hearing, and other modes of sense perception, man cannot claim any superiority over his ape-like relatives. Turning next to beings of a somewhat higher mental level, such as the apes and monkeys, we find in them a facility for making judgments in difficult situations which is often surprising. For some months I studied the behavior of a small bonnet monkey, *Pithicus sinicus,* whose intellectual capacities I attempted to gauge by a series of experiments. Lizzie's level of intelligence was considerably below that of the chimpanzee. Although curious and given to examining all sorts of objects and, whenever possible, pulling them to pieces, her attention could be focussed on any one subject for only a very short time. After she had become quite tame and would perch contentedly on my shoulder, she manifested an unconquerable dread of being seized or taken unawares. Any unusual occurrence would inspire her with instant alarm. Always watchful, she was remarkably resourceful in devising means of escape. She frequently surprised me by getting out through a half-open door which I thought was adequately guarded, and she skilfully obtained many peanuts and apples that I had not intended to give her until she had solved a particular problem. In forming good, practical judgments about means of escape, in stealthily getting food, and in making manoeuvers that involved a rapid analysis of a situation and an appropriate course of action Lizzie showed aptitude of no mean order.

Nothing in Lizzie's behaviour, however, indicated a close

approach to deliberate and formal reasoning; yet her mental adjustments to complex and varying situations were far beyond the capacity of the lower mammals. For life in a tropical forest, where she had to be ever watchful, active, and resourceful, she was an admirably adapted piece of animal mechanism, but her attempts to solve problems requiring deliberation or reflection showed that she was not equal to such tasks. When an apple was placed on a small piece of board outside her cage, she first tried to reach the apple directly, but failing in this she seized the edge of the board and with some difficulty drew it toward her and obtained the prize. When I drove a nail in the board to serve as a sort of handle she reached directly for the nail, drew in the board, and seized the apple. I then substituted a longer board, on which the apple was too far out to be reached even when the near end was pulled against the cage. Lizzie then first pulled the board in by the nail, then drew it sidewise until the apple was sufficiently near to be seized. The appropriate acts were done at the first trial, apparently as a result of inspecting the situation and judging what must be done to get the food. When an apple was placed not on the board but near it, Lizzie drew the board in repeatedly and seemed disappointed because the apple did not come also. She could clearly see that the apple was not on the board, but she did not perceive the futility of her usual performance. When given a nut inside a Mason jar, with the cover screwed on, she would bite and work at the cover, turning it this way and that until it finally came loose. Even after making numerous trials she never learned that the top could be unscrewed by continuously turning it in one direction. When food was placed in a box having a lid fastened by a button or a hook, she could open the box only by working at the fastenings with her hands and teeth until the lid happened to come open. She never clearly perceived why a button or a hook kept her from opening the box. An animal of higher intelligence would have comprehended the cause of the difficulty and performed the appropriate act without wasting so much time and effort in random biting and clawing. Lizzie was very impatient to obtain her objectives. If she could not see quickly and intuitively what course should be followed, she did not spend time in devising new methods of attack. She seemed quite incapable of exercising what we might call conscious deliberation over means of attaining ends.

I have dwelt upon Lizzie's intellectual aptitudes and limitations because they indicate a stage of mental development that is in many respects intermediate between what is found in ordinary mammals and the higher type of mentality possessed by the anthropoid apes. The recent studies of Köhler and Yerkes have added much to our knowledge of the ape mind. These studies had the great merit, as compared with older observations, of putting the animals through experimental tests in order to ascertain the character and extent of their intellectual powers. It has been shown quite clearly that the apes employ means to ends in a way that indicates a faculty of inferring what will happen if the proper conditions are fulfilled. In several experiments performed by Köhler with chimpanzees a piece of fruit was suspended beyond the reach of the animals (Fig. 2). When given boxes to mount upon, the apes quickly learned to pull them into position and climb upon them to reach the fruit. After one of the chimpanzees, Sultan, had learned to use the box the fruit was suspended still higher and two boxes were placed at his disposal some distance away from the objective. His behavior under these conditions was as follows: "Sultan drags the bigger of the two boxes towards the objective, puts it just underneath, gets up on it, and looking upward, makes ready to jump, but does not jump; gets down, seizes the other box, and, pulling it behind him, gallops about the room, making his usual noise, kicking against the walls and showing his uneasiness in every other possible way. He certainly did not seize the second box to put it on the first; it merely helps him to give vent to his temper. But all of a sudden his behavior changes completely; he stops making a noise, pulls his box from a distance right up to the other one, and stands it upright on it. He mounts the somewhat shaky structure, several times gets ready to jump, but again does not jump; the objective is still too high for this bad jumper. But he has achieved his task."

After having used two boxes Sultan and some of the other chimpanzees would pile three or more boxes one on the other. The apes blundered a great deal in their building operations; they had little conception of the conditions requisite to make the structure a stable one. Their performances were curiously like those of very young children dealing with similar problems. They exhibited a type of intelligence far below that of an adult human being. But it was intelligence far above that of an ordinary mammal.

Several observers have described how apes and monkeys use sticks or other implements in order to get objects that are otherwise out of reach. Miss Romanes, in describing the behavior of a Cebus monkey, says that "if a nut or any object he wishes to get hold of is beyond the reach of his chain, he puts out a stick to draw it toward him, or if that does not succeed he stands upright and throws a shawl back over his head, holding it by the two corners; he then throws it forward with all his strength, still holding on by the corners; thus it goes out far in front of him and covers the nut, which he then draws toward him by pulling in the shawl."

Köhler's Sultan, who had learned to use a stick to draw in bananas that were placed beyond the bars of his cage, was given two hollow pieces of bamboo, one of which would fit into the other. Food was placed outside his cage beyond the reach of a single stick. At first Sultan would use one stick to poke the other one nearer the food. These efforts of course proved to be fruitless. After this, according to his keeper, "Sultan first of all squats indifferently on the box, which has been left standing a little back from the railings; then he gets up, picks up the two sticks, sits down again on the box and plays carelessly with them. While doing this, it happens that he finds himself holding one rod in either hand in such a way that they lie in a straight line (Fig. 3); he pushes the thinner one a little way into the opening of the thicker, jumps up and is already on the run toward the railings, to which he has up to now half turned his back, and begins to draw a banana toward him with the double stick."

Sultan did not try to join two large pieces of bamboo together, but he sometimes tried to chew off a part of the end of a piece of wood that was too large to enter the hollow of a piece of bamboo and by forcing the pieces of wood and bamboo together made a jointed stick that he could use. To a certain extent, then, Sultan was not only a tool-using animal but a tool-making animal.

As Yerkes remarks, "Sharper contrast it would be difficult to imagine than that between the relatively blind and seemingly purposeless trial-and-error effort that has been described by Thorndike as typical for the cat when it faces novel problems and the definitely directed and apparently thoughtful behavior of the chimpanzee. The great apes exhibit ideational behavior; they act with insight. It remains for further patient, critical research to analyze this behavior more adequately and to compare it with our own action under identical conditions."

One of the striking characteristics of apes and monkeys is the time and effort they spend in mere play. A cow, when her hunger is satisfied, is content to rest indefinitely. But not so the ape. He must be actively exploring, romping with his fellows, climbing, swinging, or pulling things to pieces. In this way he learns much about the properties of things in his environment. His ability to use his hands in manipulating objects gives him a great advantage over the lower mammals in adding to his store of knowledge. Along with his better structural equipment he is endowed with a strong natural curiosity, which is not confined merely to things that immediately affect his welfare but manifests itself in a sort of pure intellectual interest in objects *per se.* As the ape grows older his playfulness and his spontaneous curiosity gradually diminish, and he becomes more stolid, inactive, and incurious, like so many uneducated human beings.

Chimpanzees express joy and satisfaction by smiles and laughter, especially in play and when they are tickled or given a favorite food. They often show sympathy and affection in very human ways. Madam Abreu, in describing her efforts to catch one of her chimpanzees that had escaped and taken refuge in a mango tree, writes: "I went to the tree, and speaking to him pretended that I was injured in the arm and suffering. Immediately, on seeing that I was in trouble, he jumped from the tree and coming to me held my arm and kissed it strongly. And so we were able to catch him."

"Impressive indeed," says Dr. Yerkes, "is the thoughtfulness of the ordinarily care-free and irresponsible little chimpanzee for ill or injured companions. In the Abreu collection there was for a while opportunity to observe the social relations of three individuals whose age certainly was not above five years. In the same cage were a little male and two females, one of the latter mortally ill. She was so ill that much of the time she lay on the floor of the cage in the sunlight, listless and apathetic. There was excellent opportunity to observe the attitude of her lively companions toward this helpless invalid. In all their boisterous play they scrupulously avoiding disturbing her, and, in fact, seldom touched her as they climbed, jumped, or ran about the cage. Now and then one or the other would go to her and touch her gently or caress her; or again one of them, fatigued or worsted in some game, would obviously seek refuge and respite by going close to her. In this position safety from disturbance was assured. A certain solicitude, sympathy, and pity, as well as almost human expression of consideration were thus manifested by these little creatures."

Chimpanzees are eminently social animals and are quick to resent an injury done to one of their number. (Fig. 4.) When one of them is punished "the whole group," says Köhler, "sets up a howl, as if with one voice." The excitement thus expressed has nothing of fear in it, and the group does not run away. On the contrary, they try to get to the place of punishment, even if they are separated from it by a railing. The lightest form of punishment, such as pulling the ear of the offender or a playful pretence at punishment, often stirred single members of a group to much more decisive action. "It was, in particular," says Köhler, "little weak Konsul, who would run up excitedly, and, in the way little chimpanzees have of expressing their wishes, with a pleading countenance, stretch out his arm to the punisher, if the ape was still being punished, try to hold one arm tight, and finally, with exasperated gestures, start hitting out at the big man!" It is in fact dangerous to punish an ape among a number of adult animals. Even a slight, complaining sound may bring the whole pack in an angry assault upon the offending man. It matters little what the cause of the outcry. Even a good-tempered and affectionate chimpanzee may become suddenly infected with the epidemic of rage that seizes the group, and attack one with whom it had been playing a few moments before. Mob psychology among chimpanzees is only a little more impulsive and unreasoning than it is among ignorant human beings.

The desire of a chimpanzee to be a member of a group is inordinately strong. If isolated, he is very unhappy. Some fall a prey to fears; others cry, scream, and rage violently until overcome by exhaustion. If a chimpanzee is confined alone in a cage surrounded by his comrades, it often happens, says Köhler, "that if it is only possible for them to get near the prisoner's cage, one or other of the animals will rush to it and put his arms round him through the bars. But he has to howl and cry for this affection to be shown him; as soon as he is quiet, the rest of them do not worry."

The highly social and sympathetic nature of chimpanzees often leads them to form strong attachments to their keeper. They are quick to sense the emotional attitudes of their human companions and to guide their conduct accordingly. They have their favorites among human beings, often for reasons which appear quite capricious, and they are equally capricious in exhibiting strong dislikes to individual animals as well as persons. They wish to feel that they enjoy the favor of their keeper, and they become very jealous of his attentions to other animals. The pain felt by chimpanzees on account of the master's disfavor is well illustrated by the following story told by Köhler:

"When I had been in Tenerife a few weeks only, I noticed, whilst feeding the squatting animals, pressed up close to me, that a little female, at other times quite well-behaved, was snatching the food out of the hand of a weaker animal, and as she persisted in this, I gave her a little rap. The little creature, which I had punished for the first time, shrank back, uttered one or two heart-broken wails, as she stared at me horror-struck, while her lips were pouted more than ever. The next moment she had thrown her arms round my neck, quite beside herself, and was only comforted by degrees, when I stroked her. This need, here expressed, for forgiveness, is a phenomenon frequently to be observed in the emotional life of chimpanzees. Even animals who at first when they have been punished, boil with rage, throw one glances full of hate, and will not take a mouthful of food from a human being, when one comes again after a time, will press up close, with eager bearing, to which a quick rhythmic breathing and pulling open of the eyes is added; or else will give a sob of relief, press one's fingers affectionately to their lips and make other apish protests of friendship."

It appears evident that the little ape was concerned, not so much over her punishment, as over the fact that her master could bring himself to punish her at all. Like an affectionate and sensitive child, she felt keenly the estrangement which the punishment implied, and was satisfied only when cordial relations were again established.

A lack of space forbids a description of the many little ways in which the behavior of the higher apes resembles that of human beings. I can only mention their approach to dancing, their modes of beckoning to their comrades and of conveying by movements their meaning as to what they want others to do, their extraction of thorns and splinters from their own and others' bodies, their modes of exhibiting affection, their natural hostility to strangers, and their gradual adoption of strangers into their group as they become better acquainted. For fuller information on these topics the reader may consult the works of Köhler and Yerkes, from which I have quoted.

I cannot, however, leave this topic without a few words on the family life of the apes and the care of parents for their offspring. Information on these topics is very meagre, because it is only very rarely that the larger apes have been bred in captivity. It has recently been established that the menstrual periods in female chimpanzees occur about every thirty days, and that the period of gestation is nine months—a fact ascertained from the birth of a young chimpanzee in the Abreu collection at Quinta Palatino, Cuba.

This baby chimpanzee was observed, soon after birth, in the lap of its mother, by whom it was cleaned and dried. It was quite devoid of hair except on the head. Lactation began on the second day, and the baby was nursed for several months. Incisor teeth appeared when the baby was two months old, and some of the molars developed during the next month. The mother was very solicitous for the welfare of her offspring. The father, who was somewhat morose, would often frighten the baby during his fits of temper, but was never known to molest it. Other male chimpanzees have been described as treating their young with gentleness and as playing with them.

The offspring of apes, as of most mammals, man included, receive more care and attention from the mother than from the father. Among the higher apes an adult male is often seen with one or a few adult females and a small group of younger individuals. The young cling tightly to the hair of the mother's body, and when they are old enough to run about they quickly rush to the mother whenever they are alarmed. The mother chimpanzee appears to teach her offspring to walk by taking them by the hand and guiding their steps. The babies are unable to walk alone until they are several months old.

Young apes are sometimes disciplined by their parents, and they are generally obedient to parental calls. When ill, they become objects of increased solicitude. Captain Crow tells of a small monkey that became sick during a voyage. "It had always been a favorite with the other monkeys, who seemed to regard it as the last born and pet of the family; and they granted it many indulgences which they seldom conceded to one another. . . . From the moment it was taken ill their attention and care of it redoubled; and it was truly affecting and interesting to see with what anxiety and tenderness they tended and nursed the little creature. A struggle often ensued among them for priority in those offices of affection; and some would steal one thing and some another, which they would carry to it untasted, however tempting the bit might be to their own palates. Then they would take it up gently in their forepaws, hug it to their breasts, and cry over it as a fond mother would over her suffering child."

The grief of monkeys and apes over the loss of their young has often been commented upon. After a young ape has died it is often difficult to remove the body, because the mother refuses to give it up even after it has begun to decay. The chimpanzee previously referred to, which had given birth to a baby in captivity, bore a second one about three years afterward, but it soon died. The mother would not allow it to be removed, so Madam Abreu contrived to slip a cord around the baby's neck and when the mother's attention was averted, an attendant quickly jerked the body out of the cage. The monkey, Madam Abreu states, "cried and cried, and I did my best to console her."

Young apes as a rule take very kindly to human beings. Mr. Sheak states that he had "seen a young chimpanzee, taken from the shipping box in which he came to America, throw his arms about the neck of a man he had never seen before and hug him affectionately. I had once a little fellow who would snuggle up to me, then take my arm and put it about him." Throwing the arms about the neck or shoulders of another individual seems to be a natural and possibly instinctive mode of greeting among chimpanzees. Mr. Sheak, in describing the behavior of a tired chimpanzee when she observed her master getting out her sleeping box, states that "she gave forth two or three long-drawn-out notes, followed by sharp, quick, truncated barks of delight, rushed to her master and hugged him frantically, turned to me and hugged me till she almost choked me, then hurried over to a negro at the end of the stage and hugged him too." One seldom observes such exuberance of gratitude even in human beings.

Youth is the period in which apes are most companionable and attractive. As they get older and life grows more serious, their disposition is likely to become none too angelic. And the strength of these animals renders it unsafe to take chances with the uncertainty of their temper. There are, however, as many varieties of temperament among them as there are among people, and many adult apes remain safe and devoted companions even when they grow old. Chimpanzees especially are very emotional animals, and they habitually give free rein to their impulses, whether of affection or of pugnacity. Like children, they have not mastered the arts of inhibition and dissimulation, and they are therefore all the more interesting and instructive to the student of human nature. As we learn more of the ways of these creatures, it becomes more apparent to us not only that we are very much like them but that they are very much like us.

SAMUEL JACKSON HOLMES

A PIGEON DOES IT HIS WAY

They may not have as many young as rabbits, but it only takes eight days to lay the egg, eighteen more to hatch and maybe three weeks to be thrown out of the nest. But first the young have to baby sit a couple of eggs during the day time while the folks party it up with the neighbors. Around seven months later these two young birds are able to mate as brother or sister or go their separate way and meet new company, and begin raising young.

Pigeons, just like any other birds upon contact, mate instantly. They kiss and cuddle around, then they drag their tails and in a few minutes he mounts her and the tails move to one side and in a fraction their organs touch and that's it; they parade a few seconds with their tails dragging and that's their convulsive sex orgy. From this point on they are one. Maybe a few chippies and some are a little gay along with more lesbians - this is kind of strange. Lesbian of or relating to Lesbos - the reputed homosexual band associated with Sappho of Lesbos. Sappho lived around 600 b.c. She was born in Greece and was always surrounded by her "gang" who protected her famously. She was considered one of the great Grecian poets; most of her romantic works have been lost.

BILL HUTH

GOD DID IT ALL!

Dr. H. de Borlodot, delegate from the Catholic University of Lorraine to the Darwinian Centenary at Cambridge, said: "It is no exaggeration to say that in showing us a creation more grandiose than we had ever suspected it, Charles Darwin completed the work of Isaac Newton; because for all those whose ears are not incapable of hearing, Darwin was the interpreter of the organic world, just as Newton was the voice from heaven come to tell us of the glory of the Creator and to proclaim that the Universe is a work truly worthy of His hand."

THE EVOLUTION OF THE BRAIN

IN ancient times, before paper or good parchment was easy to obtain, writers used the leaves of old manuscripts again and again after attempting to erase the writing already inscribed on them. On examining one of these palimpsests, as they are called, the modern student may be able to read sufficient of its partly erased writing to decipher the story it told. The human body is a palimpsest, which has preserved in its every part the records of its inheritance, so that anyone who carefully studies its structure and compares it with the bodies of other living creatures can read the history of man written in imperishable symbols in its very texture. Seen in this light the structure of every bone and muscle, the arrangement of the blood vessels and nerves, the constitution of the organs of body, indeed, the nature of every tissue of the organism, proclaim the fact that, wonderfully as each and every part seems to have been designed to perform its particular function, it is really a readaptation of an organ or tissue that may originally have served a useful purpose very different from that which it serves now. Like the ancient palimpsests, the human body has been inscribed again and again with new devices that have only partly obscured the more ancient writings. Let us take two illustrations:

1. At a certain stage in the normal development of a

human infant a real tail, complete, with all the muscles for wagging it, is formed; but after two or three weeks it begins to dwindle, and it finally disappears. Some of its muscles also atrophy; others are put to new purposes. No longer having any use as tail-movers, once the tail had vanished, they became converted into muscles that help to support and control certain organs of the body. Similar transformations can be found in every part of the human body; an organ or tissue that was originally developed for one purpose becomes modified to serve a totally different purpose. These statements about the tail are not theories or hypotheses, they are simple statements of fact, which any one can confirm by looking at a human embryo that has reached the third month of its development or at photographs of the embryo at that stage, which can be studied in any text-book of anatomy or embryology. The human embryo is at this stage so nearly identical with that of the monkey, dog, and pig at corresponding stages that only those who have expert knowledge can distinguish one from another. In fact, in many medical schools students examine the embryos of pigs to acquire a practical knowledge of the development of man.

2. In some animals that live in trees there is a peculiar muscle in the fore limb, or arm, which plays a part in the acrobratic feat of swinging from branch to branch. In the human arm this muscle is commonly missing, but it is sometimes found as a small and apparently useless vestige, a band of fibrous tissue representing a muscle that was a part of the bodies of our arboreal ancestors.

Neither of these illustrations is unique. The structure of any and every part of the human body tells the same sort of story of its history and affords the most unquestionable proof of the reality of man's ancient lineage and of the immense antiquity of his pedigree. The student who is searching for the truth and is competent to appreciate the significance of the facts revealed in the great adventure of exploring the structure of the human body cannot fail to discover that he himself is carrying about with him, inscribed in the very texture of his body, the record of his ancestry and of an inheritance that links him to all other living creatures.

It is often contended that such an interpretation of the evidence is merely a theory, or even nothing better than a mere working hypothesis. I want to assure my readers that such statements are very misleading—that they are actually evasions of the truth. Man's kinship with other living creatures is established by evidence afforded by his own structure, by the mode of development of his body, by the mode of action of his every tissue. We can clearly see, in the most concrete application of the term, a blood relationship.

For special consideration I have selected one particular organ of the body, the brain, because it raises the problem of the evolution, not merely of man's physical body, but of his mind, which, after all, is his most distinctive attribute. By virtue of his mental endowment man enjoys a wide vision of the world in which he lives and a high appreciation of its beauties. This endowment confers upon him powers of insight and foresight that are denied to all other living creatures. By means of speech, which the human brain makes possible, he is able to share his knowledge with his fellows, to learn from them, and to hand on the results of his accumulated experience from one generation to another.

Remembering what the human mind has achieved, the wide range of thought it has attained, the feeling for truth and beauty it has cultivated, the wonderful institutions it has created, the flights of constructive imagination it has expressed in literature and science in interpreting the meaning of human experience and the natural laws of the universe, and the thousand and one ways by which it has put into the hands of mankind the means of adaptation to the changing conditions of existence, we would seem to have some excuse for regarding men, endowed with such unique powers of intellect and sentiment, as beings fundamentally different from all other living creatures. Hence it is not surprising that the suggestion has found expression, even among such confirmed believers in evolution as, for example, Darwin's famous collaborator Wallace, that the mind is a distinctively human attribute, something that is lacking in other animals, the possession of which by man puts him in a class by himself. But no one who has made a companion of a dog and appreciates the reality and depth of his feelings and emotions, the knowledge he acquires by experience, and the sympathy and intelligence he displays in his behavior, can deny that the dog also has a mind. Though his aptitude to learn and to understand is infinitely less than that of a human child, though he seems unable clearly to anticipate what is going to happen and lacks the means of sharing knowledge that speech confers upon mankind, no one can deny that the dog is endowed with intelligence, which differs from man's intelligence not so much in its essential qualities as in its degrees—in the range of the understanding which it confers.

However, I shall not here try to define what the mind is or to discuss the question of the reality of animal intelligence. My aim is rather to call attention to the knowledge we have acquired of the instruments through which the mind expresses itself and manifests its wonderful versatility.

Every part of the human body is in a sense an instrument of the mind, a mechanism whereby its purpose can find expression—the legs that carry us, the hands that perform an infinite variety of delicately adjusted actions in the service of the will, the eyes that see, the ears that hear, the hands that feel—these and scores of other complicated pieces of mechanism in the body are surely the mind's instruments, such as in many other living creatures perform essentially the same functions that they serve in man. But almost every organ in the body plays a part in determining the appetites and desires, the feelings and the thoughts. In ancient times the Bible gave expression to the views then current among men and attributed such influences to "the reins and the heart" and to the bowels that were said "to yearn." Modern science has revealed with greater precision the part each organ plays (by means of its nervous connections as well as by the "chemical messengers" or hormones it discharges into the blood stream) in stimulating the dominant appetites and affecting our feelings and emotions—in fact, in shaping our behaviour.

I have mentioned these conceptions merely to emphasize the fact that no one organ or part of the body can be regarded exclusively as the organ of the mind, seeing that each and all, in their several fashions, may serve as instruments in exciting or expressing human behaviour. But I want to direct particular attention to the organ that plays the dominant part in our mental life—the organ whereby we are made aware of the sensory experiences that we call sight, hearing, smell, taste, and touch and all the other varieties of sensation, as well as of the feelings, appetites, and sentiments. This organ, however, controls the complicated reactions that find expression in behaviour. I do not intend to discuss the nature of the relationship between the activities of the brain and the phenomena of mind; my purpose is to call attention only to certain well-recognised facts and to discuss their meaning. We know that damage inflicted by disease or injury upon certain parts of the brain affects the mind in different ways, ranging widely from purely sensory and motor changes, such as blindness or the loss of the voluntary control of certain movements, to the most profound interference with understanding and personality. If, then, the manifestations of intelligence are in some way linked to certain bodily structures, some at least of which can be identified and examined, it becomes a matter of special interest to discover wherein the human brain differs from that of other living creatures, between whose intellectual powers and those of man there is so vast a chasm.

For more than a century anatomists have been making detailed comparisons between the brains of men and those of other creatures known as mammals, which resemble mankind in being provided with hair and with milk-producing glands for feeding their infants. It was hoped in this way to discover the secret of man's peculiar intellectual powers. Although a common and very distinctive plan of architecture is found in the brains of different mammals, an examination of their size, form, and structural details reveals a wide range of variation. The question arises whether it is possible to correlate these contrasts in the brain with the variant capabilities of the respective animals and to discover any outstanding distinctive features in the human brain that are at all commensurate with man's intellectual supremacy. A century ago, in order to explain what in those days were termed the distinctive "faculties" of man's mind, investigators plunged into this inquiry with the conviction that the human brain must contain certain organs of structures peculiar to itself. One anatomist after another put forward the claim that he had found in a certain structure evidence to substantiate man's intellectual preëminence, only to provoke someone else to discover the same structure in some non-human brain. Scores of such futile claims have been made decade after decade. But at last we have learned that there is no distinctive structure in the human brain; there is no tissue or formation that is not found in apparently as highly differentiated form in the brains of certain non-human animals. What then, it will be asked, is the real difference between the human organ of mind and that of the ape, which approaches man's brain most nearly in form?

If the brains of a series of mammals are compared, differences in shape and pattern at once become obvious. The anatomist can discriminate between them as easily as he can recognize the animals themselves. For example, compare the brain of a lion with that of a gorilla (Figs. 1 and 2). In both a great body of tissue (cerebral cortex known as the neopallium) is built up above the smaller mass that formed the older type of brain, a mass that was conjoined mainly with the sense of smell and that is relatively larger and more potent in the lion than in the ape (compare Figs. 1 and 2). The neopallium is the receptive organ of such senses as vision, hearing, and touch; it is the instrument whereby the information brought into the brain by these special senses can be blended and recorded so as to be recalled in memory and to influence the movements and behaviour of the whole organism.

By comparing the brains of the lion and the gorilla, creatures of roughly the same bulk, we may see that, although in the lion the sense of smell is more strongly represented than in the gorilla, the other senses (expressed in the neopallium) are much more strongly represented in the ape (Fig. 2). Moreover, the plan formed by the folding of the neopallium in the lion (Fig. 1) seems to be so utterly unlike that of the ape that anatomists are still disputing whether

Fig. 1

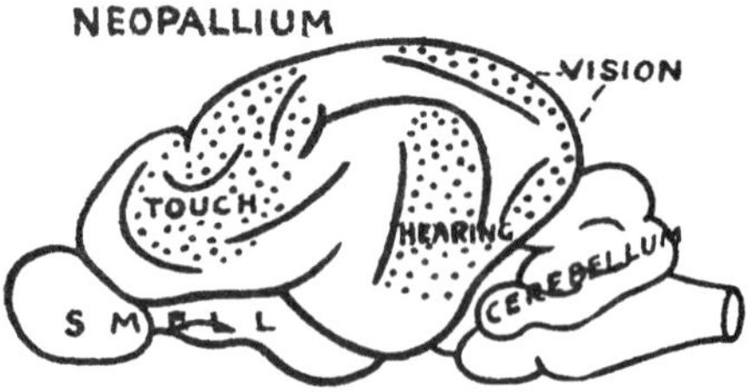

Fig. 2

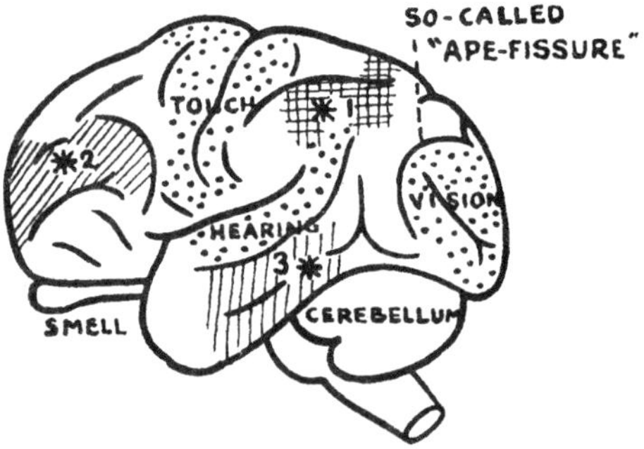

Fig. 3

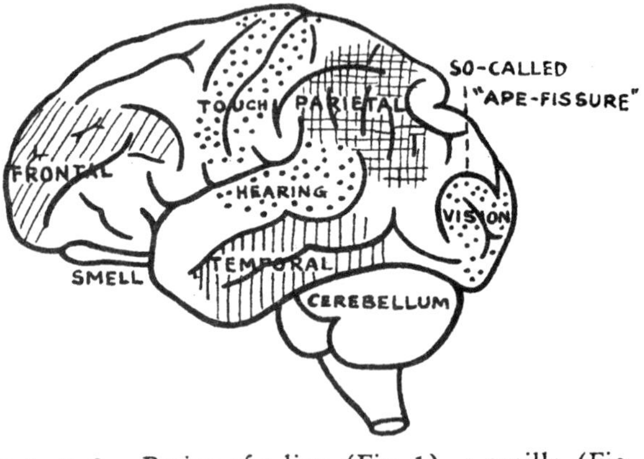

FIGS. 1, 2, 3.—Brains of a lion (Fig. 1), a gorilla (Fig. 2), and a human being (Fig. 3), seen from the left side.

there is any homology. Bearing in mind these profound differences in pattern it comes as something akin to a shock to discover that the general pattern of the brain of a gorilla (Fig. 2) is essentially identical with that of the brain of a man (Fig. 3). In spite of the fact that the exceptionally small and very primitive human brain represented in Fig. 3 is double the bulk of the ape's (1,000 grammes and 500 grammes, respectively), the form and pattern of the gorilla's brain so closely reproduce those of the human brain that it is easy to recognise corresponding areas in the two. By comparing these areas the further fact emerges that the parts that receive the impulses of vision, hearing, and touch are almost, if not quite, as extensive in the brain of the ape as in the brain of man. In fact, the only noteworthy differences are due to the enormous expansion in the human brain of three areas, the representatives of which in the gorilla's brain (distinguished by stars in Fig. 2) are comparatively insignificant. Studies of the effects of diseases or injury upon the brain has shown that damage to these particular areas of man's brain has the most profound effects upon intelligence and personality. Moreover, one who studies the microscopic structure of these areas that we know to be concerned with the expression of man's highest mental powers, the instruments of his distinctive intelligence and personality, can detect no difference in structure or quality between them and their diminutive representatives in the brain of the ape.

The accompanying diagrams show the profound contrast between the patterns of the brains of the lion and the gorilla and throw into relief the remarkable identity of structure of the brains of the gorilla and of man, although the human brain is double the volume of the gorilla's. The human brain chosen for this purpose is exceptionally small (about two-thirds the average size) and is primitive in type. The difference in size is confined almost wholly to three areas, and it is profoundly significant that the areas which reveal the enormous expansion in the human brain are precisely those that attain their maturity latest in the human child (Fig. 4). The more precocious, such as 1 (the area con-

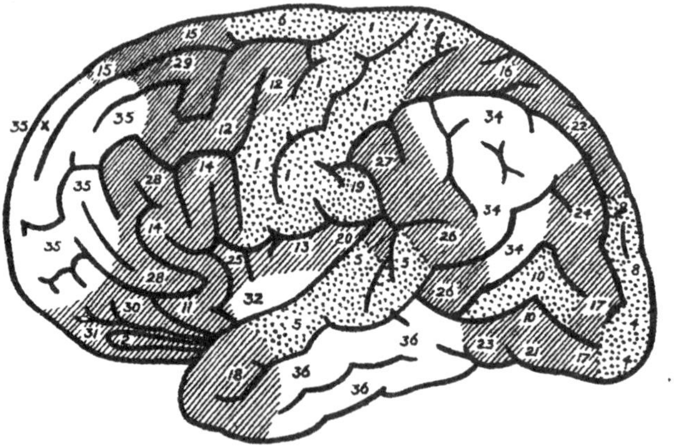

FIG. 4.—Brain of a child.

The numbers show the order in which certain areas of the human brain are perfected. Reproduction of a chart made by Professor Flechsig.

cerned with the sense of touch), 4 (the visual receptive area), and 5 (the acoustic area) are just as well developed in the ape as in man. The intermediate areas (6 to 30) are also moderately large in the ape. But the latest areas to mature (34, 35 and 36) are enormously bigger in the human brain. The only contrast between the human and the simian brain is that certain areas in the brain of man are enormously bigger than the corresponding areas in the brain of the ape (Figs. 2 and 3). The structure of these corresponding parts

—perhaps the most amazingly complex pieces of living machinery—is so similar that even the most experienced anatomist is unable to distinguish between them.

The physical instruments that are the sources of man's highest mental qualities are thus represented in the brain of the ape. Their construction in both is identical, but in the ape they are very much smaller. The difference between the brain of a man and the brain of an ape are not qualitative but quantitative. The ape has the germs of the mental powers that are man's supreme distinction. This conclusion has recently been confirmed by the careful observations of Professor Yerkes of Johns Hopkins University and Professor Köhler of Berlin, who have devoted years to the study of the chimpanzee's behaviour. When serious consideration is given to the identity of structure in the brain of man and the brain of the man-like ape—even though the ape's brain may be but half or a third of the bulk of the human brain—the only conception that affords a credible explanation of the resemblance is that ape and man had their origin in a common, even if very remote ancestor.

But if it be admitted that men and apes are derived from a common source, though the apes have neglected to develop their possibilities as men have done, the implication is that the work accomplished by man's brain, which finds expression in the human mind and personality, must necessarily be of the same kind that a brain of simian type is capable of doing and that man has been evolved from some lower type.

It must not be forgotten that on the most conservative estimate it is much more than a million years since the ancestors of man and of the apes parted company from their common parents. The apes gradually lost the power to develop further the brain and the mind in the way that the ancestors of man were able to do, because the apes became adjusted to particular modes of life, so that their brains and hands, and in fact their entire bodies, lost that power of adaptation to new or changing conditions which the ancestors of man retained. Hence it is altogether unlikely that in the future any ape can be transformed into a man. The ape's thumb is already so atrophied that it can never regain its adaptability, and without adaptable hands, which are the instruments for applying knowledge and for developing skill, the brain cannot progress in the way necessary to attain the human type of intelligence.

At one time it was generally believed, as I have already remarked, that the ape's brain has distinctive features, which were lacking in the human brain. One of these was regarded as so eminently characteristic of monkeys and apes (Fig. 2), that it was called "the ape-fissure," or more usually Affenspalte, the German equivalent of this expression. But all these features, and in particular the so-called "ape-fissure," have now been found in the human brain (Fig. 3). The fact that they were formerly believed to be so peculiarly distinctive of the apes assumes special significance now that their presence has been demonstrated in the human brain; for they become further tokens of the close affinity between man and the ape—labels, so to speak, to force us to recognise in the organ that is in a sense the physical expression of man's intellectual supremacy the evidence establishing its community of origin with the brain of the ape.

Nor must we restrict the brain's activities to the regulation of the bodily functions and the manifestation of intelligence. The brain is the organ that controls behaviour. As Charles Darwin said, more than fifty years ago: "A moral being is one who is capable of comparing his past and future actions or motives, and of approving or disapproving of them. We have no reason to suppose that any of the lower animals have this capacity; therefore, when a Newfoundland dog drags a child out of the water or a monkey faces danger to rescue its comrade or takes charge of an orphan monkey we do not call its conduct moral." If this be admitted it follows that one of the essential conditions for the display of moral qualities depends upon the integrity of certain parts of the brain, without which it would not be possible to recall the past or to speculate about the future. Memory and foresight do not, of course, confer moral qualities, but they do represent conditions essential for the display of such humanitarian attributes.

Only during the last century has any accurate information been acquired as to which parts of the brain are concerned with the intellectual functions. Even at the present moment the terrible effects of the damage inflicted upon the brains of patients suffering from what is popularly known as "sleeping sickness" are opening new vistas of knowledge as to the parts of the brain upon whose activities personality, the sentiments and emotions, muscular skill, and intellectual and moral capabilities depend. In scores of patients, during the last five years, physicians have witnessed the most profound changes in character and morals when this insidious disease has destroyed certain small areas of the brain.

Three centuries before the beginning of the Christian era some of the wise men of Greece already recognised in the brain the real organ of the mind; yet it was reserved for modern times to confirm the accuracy of this early knowledge and to extend it. In olden times the seat of the understanding was placed in the heart, as every reader of the Old Testament must be aware, although certain passages in the

Bible suggest that sometimes the kidneys were regarded as the organs of mind, or both the kidneys and the heart, as in the New Testament (Revelation, II, 23), where the Son of God says: "I am he that searcheth the reins and the heart."

Even the Psalmist who wrote the oft-quoted verses (Psalm CXXXIX, 13-16), which have become the special hymn of the anatomist and embryologist, begins with the suggestion that the kidneys are the seat of the will:

Thou hast possessed my reins:
Thou hast covered me in my mother's womb.
I will praise thee: for I am fearfully and wonderfully made: marvellous are thy works,
And that my soul knoweth right well.
My substance was not hid from thee
When I was made in secret
And curiously wrought in the lowermost parts of the earth.
Thine eyes did see my substance yet being imperfect,
And in thy book all my members were written which in continuance were fashioned
When as yet there was none of them.

But of all these works none is so marvellous, so fearfully and wonderfully made as the brain, which confers upon man the supreme distinction of his high powers of intelligence.

The brain affords evidence in corroboration of man's origin from an ancestor common to man and ape that is too exact and impressive to admit of any doubt as to its significance. By demonstrating that the structures concerned in the highest expressions of human intelligence are already present in the ape's brain, even if they are very diminutive, the study of the brain adds strength to the conviction that the mind as well as the brain has been evolved. The fact that the Bible in various places assigns the chief seat of understanding sometimes to the heart, at other times to the kidneys, or both heart and kidneys, has not been allowed to interfere with the general recognition of the preëminence of the brain as the organ of mind. Why, then, should any difficulty be raised in opposition to the patent fact that the brain itself "in continuance was fashioned," in strict accordance with an inherited plan, which is common also to that of our nearest living relations in the animal world?

I have here called attention to the fact that the rapid development of our knowledge of the human brain and of the effects of injury of disease to different parts of it has made it possible for us to identify the structures whose activities find expression as mind and personality. In the brains of other living creatures corresponding structures can be detected, which conform in every respect except size to those areas which in man we have recognized as the special instruments of the mind. The resemblance of the brain of some creatures, like the chimpanzee and gorilla, to the brain of man is much closer than that of either to the brain of any other animal. The only reasonable and satisfying explanation of such close resemblances, both in structure and in function, is the inference that (a) these other creatures have the undeveloped germs of a mind similar in kind to man's (one, however, that has definitely lost the power of significant development or further progress of the kind distinctive of man's immediate ancestors), and (b) that both the brain and the mind of man are the results of a long process of development from ancestors common to those of other living creatures having brains of the same essential type.

G. ELLIOT SMITH

YAHOO!

Lyell, in a letter to John Hershel, in 1830, wrote: "When I first came to the notion . . . of a succession of extinction of species, and the creation of new ones, going on perpetually now, and through an indefinite period in the past, and to continue for ages to come, all in accommodation to the changes which must continue in the inanimate and habitable earth, the idea struck me as the grandest which I had ever conceived, so far as regards the attributes of the Presiding Mind."

PROPHET OR NOT?

Prophet, n. - Synonym (one or two more words or expressions of the same language that have the same or nearly the same meaning in some or all senses). So here we go, this is what they can call a prophet: seer, seeress, oracle, soothsayer, clairvoyant, wizard, augur, sibyl, sorcerer predictor, forecaster, prognosticator, diviner, haruspex, evocator, medium, witch, palmist (that's what they call a bust-out crap dealer), tea-leaf reader, reader of the future, weather forecaster, meteorologist, ovate, bard, druid, vats sacra (Latin), magus, astrologer, horoscopist. In my own mind I always thought of a prophet being just a prophet. I guess your only a prophet when you say something to the tip, (a bunch of suckers in hustler talk) they want to hear. Otherwise they got you pretty well covered for a choice words to explain really what u might be. U know I've been wasting a lot of time spelling u, you. I think Arron had a little soothsayer or wizard in his make-up. I think his brother Moses had a hard time, at times, keeping Arron straight with the mob, or tip if you want. I believe Paul was the ramrod behind the deciphers, or Apostles, which evers (another short cut) best for you, and he pulled a little prophetic hoky-poky too. Maybe he didn't eat enough and that caused it. They say the more you want to see; the less you eat. U can imagine what forty days would be like!

BILL HUTH

EVERY DAY IT IMPROVES FOR CREATION'S BENEFITS

"It is plain that neither in 'systematic theology' nor in science has the last word been said. In astronomy, in physics, in life, in space, in time, in thought, we find ourselves baffled in the face of Infinity. The Master Key that shall unlock all doors which open toward the center, no man has yet found. It too must lie within the gates of Infinity!"—David S. Jordan.

MAYBE IT IS ONE STROKE; THAT TOOK BILLIONS OF YEARS TO SWING

"Organic evolution states most emphatically that species are not fixed and unchangeable, and were not created in one sudden stroke, but that they have varied considerably and that the forms now existing have slowly developed from more primitive ancestors."—Joseph Meyer.

I'VE HAD SEVENTY THREE YEARS OF IT

"To understand what has happened, and even what will happen we have only to examine what is happening."—Buffon.

NOTH'NS MISSING

Evolutionists, Darwin included, do not say that man is descended from any existing kind of ape or monkey, but that pro-man and ape, in the dim and distant past, had a common ancestor, now extinct, that was neither man nor ape.

FRANCES MASON

PROGRESS SHOWN IN EVOLUTION

SOME seem to suppose that evolution is synonymous with change, even if the change is disorderly and chaotic; but if we look at evolution as it actually exists, whether the long-range evolution of species from species or class from class, or the short-range evolution that occurs in the individual development of each human being and each familiar animal from the egg to the adult stage, we find that one of the characteristics of evolutionary change is its orderliness. Each step in each separate evolutionary line is orderly, its significance can be fully understood only as the result of what has gone before and as the necessary prelude to what is to come after. If we turn from single lines of evolution to the evolution of life as a whole, we can ask a new question. Granted that the separate changes of evolution are orderly, can we discern one sole or main direction, or a few main directions, in the general evolution of life? Finally, if we were to find that evolution followed only one or a few main trends, can we say that these trends or directions are, in any real sense of the word, *progressive?*

The answer to the first of these two questions is definite enough. In its march through time life does follow certain main directions. This fact can be shown by actually tracing the history of animals through geological time by means of their fossil remains, by deciphering the history of the race from the summary of it presented in the history of the developing individual, and, indirectly, by comparing different animals with one another, a comparison that enables us to deduce with reasonable accuracy their family history and genealogical tree. By combining all the various lines of evidence it is possible to arrive at certain perfectly definite conclusions. In the first place, the size of animals has increased during their evolution. None of the earliest mammals were much bigger than a dog; creatures the size of a horse or a hippopotamus were unknown. The same is true of the reptiles; the giant forms appeared late in their geological history. The earliest forms of life were doubtless microscopic. Even the smallest mammal is over a million million times as large as a medium-sized amoeba.

More important than mere increase in size is increase in efficiency. No early form of life can move fast or can see or hear acutely; none has a heart or a blood-system, brain or nerves, jaws or limbs, or true head. The earliest true vertebrates have no jaws or teeth. The earliest land vertebrates could not support their bodies with their limbs. The earliest members of the horse family had limbs incapable of the speed achieved by the later horses, and the teeth of an early horse are not nearly so efficient for grinding as those of a modern horse. The brains of early mammals were barely more than half the proportional size of those of present-day mammals of the same bulk. The improvement of the efficiency of separate organs is the improvement of the tools of life; the improvement of the efficiency of the brain is the improvement of the way in which these tools can be used.

When we examine the geological history of a single group of animals, such as the mammals or the reptiles, whose past record is available to us in the fossil-bearing rocks, we find that improvement in efficiency is actually made in different directions by a number of separate evolutionary lines. The improvement of each line means improvement of efficiency at one particular job, such as improvement in one way of getting a livelihood or in planting its offspring securely in the world. For instance, the early mammals of the Secondary period (Mesozoic age) were small land animals whose fore and hind limbs were not very different from each other —four or five toes on each foot, teeth of a primitive type like those of a shrew or a hedgehog, and small brains. During the Tertiary period these creatures gave rise to a whole set of types. Some of these types, like the whale and the dolphin, are adapted to life in water; others, like the horse and the deer, are adapted to herbivorous diet and rapid running; the higher carnivores, like the lion and the wolf, are adapted to a flesh diet, which is captured by speed, strength, and skill; the bat is adapted to a life in air; the elephant survives by virtue of its huge size and formidable tusks; the mole lives by its adaptation to a burrowing life below ground; the sloth, by its adaptation to life in trees; armadilloes survive by their protective armour.

Each of these creatures represents the finished product of a line of improvement in one particular tool or method. On the other hand, each individual improvement has been made at the expense of other possible improvements. The original type was primitive but plastic ; it was capable of being altered in many possible ways. But the whale, in becoming an efficient swimmer and diver, has lost any possibility of ever being able to run or to fly; the horse, by gaining its efficient running organ in the shape of a long leg with but one toe, has lost the possibility of acquiring a hand for grasping or a clawed foot for catching prey; the elephant's bulk precludes agility; the mole, though good at burrowing, is worse off as regards the possibility of climbing trees. Such improvement is best termed *specialization,* and the complete set of divergent specializations which characterize the evolution of a whole group, such as the mammals, is called the *adaptive radiation* of the group. Biological specialization moves always in one direction only and is achieved at the expense of improvement in other directions. What is more, specialization in improving the efficiency of a physical tool, such as a limb or eye, is bound sooner or later to reach a limit. The elephant is pretty close to the limit of size which is possible or at least advantageous for a purely terrestrial animal. The speed of wild horses or antelopes is close to the greatest speed that is possible to a four-legged land animal; acuteness of vision must reach a limit owing to the impossibility of obtaining cells in the retina below a certain size; and so forth.

Thus specialization and adaptive radiation, though they increase immensely the efficiency of life as a whole and enable it to reach its greatest limits in this or that direction, are yet in a sense double-edged. In opening the door to one kind of improvement they close it to other kinds, and in the long run even turn out to be blind alleys, to which positive limits are set. We can easily recognize the limitations of specialization as a method of evolutionary improvement by considering specialization for a parasitic existence. An internal parasite, such as a tapeworm or a malarial parasite, has no need to find or to digest its own food, to move from place to place, or to detect enemies at a distance. Accordingly we find that most internal parasites have no mouth or digestive system, no means of locomotion, or very much reduced means, no well-developed sense-organs. On the other hand, parasites must be specially adapted, for instance, to resist the action of digestive juices or of protective devices in the blood of their host, and especially to enable them to get transferred from one host to another. We thus find that in respect to reproduction and life-cycle, parasites are usually much more elaborate than tree-living animals; for instance, the dog tapeworm can be transmitted to a dog only by entering the body of an animal like a rabbit, there going through a special cycle of life, and then being eaten again by a dog.

We usually say that parasites are degenerate, because we note their striking loss of organs and faculties, but they are only particular examples of specialization, with, as usual, elaboration and improvement in one direction and loss in others. The whale, in gaining blubber and tail fin, has lost hair and hind limb; the horse, in improving its middle digit to a hoof, has lost the other four digits on each foot.

But besides such one-sided specialization, there are examples of evolutionary improvement which are all-round, or balanced, and do not deprive their possessors of their precious plasticity. For instance, the change from cold-bloodedness to warm-bloodedness in vertebrates was such a change. In becoming warm-blooded, the bird or the mammal lost nothing which their reptilian ancestors possessed; they merely acquired a new and valuable piece of vital machinery, which enables them to be much more independent of the temperature of the outer world than they were before. In the same way, the reproductive methods of reptiles and birds represent a pure gain when compared with those of their fish-like and amphibian ancestors. The evolution of the protective membrane or *amnion,* which makes a water cushion round the embryo, and the other embryonic membrane or *allantois,* which enables the embryo to breathe within the egg-shell, made it possible for reptiles to be independent of water for their breeding, and so helped to open up to them vast tracts of the earth's surface which other vertebrates had until then not been able to conquer. Such changes, involving the improvement of the all-round achievements of the organism without depriving it of valuable possibilities, may properly be called biological progress. They are simply examples of specialization that is not one-sided, but balanced.

We may take one further example, which brings out the difference between the two processes. The most primitive members of the group to which we and all other backboned animals belong—forms like *Amphioxus,* for instance—have no true eye, have probably only a very slight sense of smell (certainly no nasal organ of our type), and no ear. The lower vertebrates, such as the fishes, have very efficient sight and smell but practically no sense of hearing. Both birds and mammals (in general) have acute hearing and much improved sight. Here there is a real biological advance; the efficiency of all three senses has enormously improved, and improved in a balanced way, in passing from *Amphioxus* to higher vertebrate. But in this same field we may find unbalanced improvement, one-sided specialization. The improvement in the utilization of the sense of sight, which is so obvious in the whole group of monkeys and apes and culminates in man, has been accompanied by a degeneration in the power of smell; the same has been true in many birds, which also rely almost entirely upon sight. On the other hand, the mole relies almost entirely upon touch and hearing, and its eyes have degenerated. Thus in all these forms an unbalanced improvement in one direction has led to a cutting down of faculty in another.

The main improvements of life during its evolution must obviously be improvements of the balanced type, not mere specializations, since it seems certain that no highly specialized animal or plant has ever succeeded in becoming the ancestor of a new group or type, such being the privilege of biologically balanced or generalized creatures.

The most important progressive steps in the evolutionary ascent of animal life perhaps deserve mention. Starting from the single-celled type, life made its first great advance through the aggregation of many single cells into a colony; this advance was followed by division of labour for different functions among different kinds of cells, which gave new possibilities of size and balanced specialization of function. Next came the organization of the community of cells into a two-layered creature with a mouth at one end, a stage preserved to-day in sea-anemones and their relatives. Then came the intercalation of a third layer, and the development of a centralized (though primitive) nervous system and primitive kidneys. Then the development of a blood system, a posterior opening to the digestive tube, better locomotor organs, and elaborated sense organs in a region which might properly be called a head. Leaving all but the vertebrates out of consideration for lack of space, we would next come to the enlargement of the brain, the development of a strong internal skeleton, and then to that of paired limbs. These improvements are followed by partial emancipation from the water, as in the amphibians, then total emancipation, as in the reptiles. Still later we find the attainment of the condition of constant temperature, called warm-bloodedness; the improvement of the nourishment and care of the young, both before and after birth; and the rapid improvement of memory, associative power, and animal intelligence. Finally, in man, comes the new step in brain power which we call reason—the power of generalizing, and consequently of giving names to things, and so speech, which has brought in its train the other enormously important progressive development, the possession by the human species of experience and tradition that is cumulative from generation to generation. At each of these levels some types of living beings have specialized and remain fixed to this day in their one-sided efficiency, or else they have been extinguished by more progressive types; others have remained generalized, and some of these have given birth to the progressive types which constitute the next upward step.

After this brief survey, it remains to ask whether the balanced advance we have been discussing can properly be called progress, in the usual sense of that term, or whether we have not been misleading ourselves by using a term which implies real improvement in what to us is valuable, when we should have really called it mere directive change.

When we come to consider the main steps in biological advance that are enumerated above, we find that it is possible to sum them up under a few heads. There has been on the whole a considerable increase in size; there has been improvement in the organs adapted to carrying out each type of function taken separately—organs of digestion, of locomotion, of protection, of support, of sense-perception, of reproduction; there has been improvement in the relation between these organs—that is, in the way in which the different parts of the body and their functions are correlated and coördinated; there has been improvement in the control exercised by the brain over the body as a whole, and in the quality and extent of the information received about the outside world by which this control is achieved; there has been improvement in the self-regulating capacity of the body, as is witnessed by constant temperature or constant chemical composition of blood in higher forms; there has been a decreasing reproductive waste, an increasing care for young; there has been an increase in mutual aid between individuals; there has been an increase of emotional power and of purposive action. If we now examine this list further, we find that every one of the improvements enumerated may be thought of as conferring upon the individual or the race increased power of control over environment, increased internal harmony and self-regulative capacity and consequently increased independence as regards the outer world, increased knowledge, and increased intensity and harmony of mental life.

Whether the list is considered in its first state or in its second, there are very few who will not admit that these biological improvements, which have made for survival and success in evolution, are not also improvements when judged by our human standards of value. We, too, strive for control over nature and for greater independence of outer conditions; we value harmony; we prize knowledge and all the products (when balanced) of increased intensity of emotion and will. It is therefore justifiable, since progress is a word which implies progress toward something which we men find of value, to speak of the observed movement of life that we have so far called *biological advance* as real *biological progress.*

It may be argued that this is mere reasoning in a circle; that of course, as we are ourselves products of the evolutionary process, we shall find its movement coincide with our ideas of good. This is in reality not so at all. It is *not* all kinds of evolutionary movement which we find good in this way, but only the one kind that we have defined as balanced advance. There are many other kinds of evolutionary process. There is, for instance, extinction. Whole groups of animals and plants, some of them of remarkable vigour, size, beauty of adaptation, have wholly disappeared from the face of the earth. The trilobites, the ammonites, the wonderful dinosaur group of reptiles—these are but three examples; and even to-day the process is being very rapidly continued by man, who is, often needlessly, exterminating entire species—strange creatures like the great auk, lovely ones like the passenger pigeon or the sea-otter. Extinction is in itself not a good but an evil. It can only be a good on balance, if and when it is necessary that one group or type should perish that another more advanced group should flourish.

Then, as we have seen, specialization, though sometimes we should call it good, is never an unmixed or a balanced good. Not only that, but in some of the examples considered, such as parasitism, the balance is the other way, and what seems good to us is outweighed by what seems clearly evil. If tapeworms could reason and formulate their opinions about the universe, they would have to admit that the general trend of evolution was very different in its direction from that to which they owed their being, and that on the whole the two were opposed. They would, presumably, have to adopt that philosophy or belief which characterized the Manicheans of the early Christian era. Even the products of specialization that to us is clearly on balance good, though limited—such products, for instance, as the birds—would (if they were able to think it all out) think in rather different terms. Their specialization has led to flight, to intense activity, to colour and song unrivalled among the mammals. Well might they pity earthbound, drab-coloured, hairy creatures, and maintain that activity and the conquest of the air were the highest achievements of evolving life. But they would be wrong. It is simply a matter of hard fact, which takes no account of actual human wishes or hypothetical bird wishes, that for some reason (probably their sacrifice of fore-limb for a tool of flight) the birds' brain-development has been restricted, whereas the mammals, evolving along the lemur-monkey-ape line, were able to develop brain power, which finally culminated in man, and which has made man now the dominant, most successful organism and has enabled him actually to beat the bird at its own game, producing machines more swift and tireless in the air than the birds themselves, and music compared to which even the lark's and the nightingale's songs are but naïve.

No, the only reason why we find that the direction of biological progress coincides so closely with much of our own ideas of progress and value is that man happens to be in the main stream of biological progress, not in an eddy or backwater.

The fact that there exists (among other processes of evolution) one which can rightly be styled progress, seems to me of fundamental importance to our thought. It does not in any way prove the existence of a supernatural purpose in evolution. It is merely one among several kinds of evolutionary results, no one of which is any less explicable by natural causes than any other. Paley and his school maintained with great vigor that the existence of adaptive structures closely fitted to the function they were to perform—the webbed foot of a duck, the eye of man, the stream-line form of a fish—were proof of a supernatural designer. Darwin at one stroke swept this argument away by means of his theory of natural selection. This was early recognized by all who accepted Darwinism, but it has not been so readily recognized that precisely the same arguments hold as regards biological progress: granted the existence of Variation and Natural Selection, then biological progress as well as adaptation (which is the product of specialization) must come about.[1]

[1] See essay on Progress in J. H. Huxley's Essays of a Biologist.

But the fact of biological progress does show that our ideals and efforts, our whole scheme of values, are not merely isolated flames burning in the darkness of a universe which is neutral or hostile to the effects of its working. It shows, at least as regards the course of events for the several thousand million years during which life has existed on this planet, that the cosmic forces have worked in such a way as to produce a movement that has been not only the most successful movement in evolution but that also chimes in with our sense of values and our idea of the direction in which we ourselves desire to move.

As a result of the working of the forces of Evolution, man has now become the trustee of the evolutionary process—in other words, of this world's future. It is sometimes asserted that pre-human evolution is purposeful. This, however, is a mistake. As Darwin once and for all pointed out, the adaptation of structure to function, of an animal's means to the requirements of its life, not only require no other purpose at all than the purposes of the contending animals but can be explained only as the result of the automatic working of natural forces, such as heredity, variation, and natural selection; and the same statement is true if we consider the general movement and progress of life's evolution as opposed to the particular evolution of this or that species. These natural forces are in the same category as those which cause chemical combinations, make the chick grow in the egg, or cause our digestive juices to be secreted; but purpose is something that can be defined only psychologically—meaning that a definite end is consciously held in view.

This of course is not to say that there may not be purpose in the background, behind all those apparently blind forces. But we have no means whatever of knowing whether that is so or not; and it is the first duty of anyone who wishes to think scientifically to suspend judgment on such questions in the absence of evidence.

What is definitely true is that the forces *which we can actually detect* operating in the evolution of plants and lower animals are automatic and non-conscious; whereas those operating on the human level, as we can again obviously verify for ourselves, are in part conscious and include ideals of truth, beauty, and morality. We may even say that the forces of evolution conspire to act as "a power, not ourselves, which makes for righteousness."

Our business on this planet, then, is not to worry our heads about the possible forces which *may* exist behind those which we know, but to strive to mould the known material forces of dead and living matter in accord with the spiritual ideals of value which we possess.

JULIAN SORELL HUXLEY

ECCLESIASTES 3:18-22

"Is it not the most sublime, the most stimulating conception that has ever entered human thought, this conception of progress, this new idea absolutely unknown in ancient times, a progress of which we are a part, and in which we are ourselves consciously playing a rôle of supreme importance?"—Robert Millikan.

OUR TRUST

"I do not know what I may appear to the world, but to myself I seem to have been only like a boy playing on the seashore and diverting myself now and then finding a smoother pebble or a prettier shell than ordinary, while the great ocean of truth lay all undiscovered before me."—Sir Isaac Newton.

MIND IN EVOLUTION

OPINIONS differ as to what the word evolution means or should mean. Some writers speak, for example, of the evolution of atoms, molecules, crystals; of the solar system, the Alps, the Mediterranean Sea; of plants and animals; of social institutions; of scientific thought or artistic expression. For these writers the word evolution is unrestricted in its application. But other writers are of opinion that it is better to restrict the meaning of the word to what is spoken of as the doctrine of descent in living creatures, in other words, to that which, broadly speaking, falls under the head of the origin of species.

This question has been much discussed and a good deal has been said on both sides. But one must choose one or the other. I use the word in its unrestricted sense. Under this usage one can add appropriate adjectives to qualify the noun, such as cosmic, physical, chemical, organic, mental, and social evolution. That enables one to make clear what one means.

But if we adopt the unrestricted use of the noun to express a broad, comprehensive idea we must expressly state the particular phenomena that exemplify the idea we have in mind. We seek, then, to ascertain what holds good for all natural events in so far as the concept of evolution is exemplified in them. It is not easy for me to put it clearly or for my reader to understand. But let us both do our best.

What seems to be common to all events, no matter what may be their specific character, is that which may be called *passage*. They pass from one phase or stage to the next. And the passage is accordant with some general method or plan which we can more or less definitely formulate. We should realize that even what we commonly call "things" are relatively persistent clusters of events in passage. They have been likened to eddies in the stream of events, or to a waterfall, where the water flows on but the cascade is permanent. What persists is some *state* of flowing events.

Suppose, then, that we are dealing with atoms, molecules, organisms, minds, social institutions. In all of them there is passage of events. In each there are relatively persistent states, which characterise the several members of the group —characterise each atom, or crystal, or organism, or mind. But in any given group the individual members are not all alike. Molecules are not all alike; nor are organisms or minds all alike. And we commonly say that some of them stand at a *higher* level, some of them stand at a *lower* level than others. Thus man is at a higher level and a monkey at a lower level than an ape. If we speak of the level at which any member of a group stands as its *status* then a man has a higher status than an ape, an ape has a higher status than a monkey, and a monkey a far higher status than an amoeba.

Of course, we have in some way to define what we mean by "higher" and "lower." We may say—to select *one* character—that what is more complex is higher and what is less complex is lower. Then in this respect within each group the members may be arranged in order from the lowest to the highest. And the groups also—the atoms, molecules, crystals, organisms—stand in an ascending order of status from lower to higher. Molecules have a higher status than atoms; crystals have a higher status than molecules; organisms have a higher status than anything inorganic.

Suppose, next, that we are dealing with a group of organisms, let us say plants. As we have seen, they may be arranged in an ascending order according to status. But in the course of individual development from the seed onward there is, for instance in the oak, a passage of state from the less complex acorn to the much more complex oak tree. And in the course of racial development, according to the doctrine of descent, there has been, in times long past, a passage of status from less complex species of plants to more complex species.

Now this kind of development in the individual and evolution in the race is not found in atoms, or molecules, or crystals. It is not found till the level of living creatures has been reached in the progressive advance of nature. It introduces something quite new and distinctive—what we call life—which, in technical phrase, "differentiates" organic from inorganic evolution. This makes a difference in the course of events. To indicate other differences the adjectives atomic, molecular, chemical, mental, and so on are used. But the *noun* "evolution" is here invariably used to mark something which is common to all of them.

After what has thus been said—and necessarily said very briefly—we are now, I think, in a position to state what is common to all of them. Laying stress on passage of states and of status, we can give a pretty clear meaning to unrestricted evolution. It means upward passage from lower to higher, no matter what particular form this passage may assume in this or that kind of progress. The emphasis on evolution as thus defined is therefore on its *upward* passage.

Thus what from our point of view is essential to the idea of evolution is upward passage by progressive steps (sometimes very little steps, sometimes big jumps) along definite, recognisable lines of advance, with continuity of progress from lower to higher. And of evolution in this sense there is evidence in molecules, in organisms, in minds, and in social institutions.

No doubt, when we come down to adjectival details, we shall find special features that are distinctive of each group, and some difficulty may still be felt in defining *advance.* I have spoken of advance to what is higher; and to illustrate what may be taken as a criterion of higher, I selected complexity. But this is not the only criterion, perhaps not the most important criterion. For, in the upper reaches of evolution, what is higher may be higher in *quality.* Thus one man's treatment of a subject may be higher than that of another man not in complexity but in what we commonly speak of as "quality." A little dinner may be higher in quality than an elaborate banquet. This distinction may be hard to define, but most people will understand what is meant.

If, with a little attentive thought, one has grasped this idea or concept of evolution as *upward and progressive advance,* the next thing to realise is that, throughout nature, including human nature, there is by no means always progressive advance. In every field of inquiry we find abundant evidence of that which is the very opposite of evolution and is sometimes called "degeneration" or "devolution." I shall speak of it as dissolution. Evolution is progress, dissolution is regress. What we have now to grasp is that we find throughout nature not only upward passage from lower to higher but downward passage from higher to lower—sometimes one, sometimes the other; sometimes both side by side. Where we find both we usually find a balance in favour of one or the other. We have often to deal with an intricate profit and loss account.

In our own bodies, for example, the tissues are not only in process of up-building but also in process of down-breaking. During adult life some sort of balance is maintained on the credit and the debit side of the account. In the healthy child there is a balance of profit through evolutionary upbuilding. In old age there is more loss than gain. In what we speak of as "senile decay" dissolution, through degeneration of the bodily tissues, entails an increasingly adverse balance.

This distinction between building upward under evolution, and breaking down under dissolution is of very great importance. Surely, wherever we look we find not only progress but regress; we find not only building up but breaking down. That which is progressively built up under evolution has that mark of the higher which stamps it as a fuller and richer *whole* with substantial unity. That which ultimately results from dissolution is the scattering of the components which went together to constitute *that* whole. Much modern work on the atom illustrates dissolution; of atoms in process of evolution there is now little evidence. But many of us believe that it has taken place in the past and may still occur. And what about social life? Here we find abundant evidence of evolution in progress. But is there no evidence of dissolution in regress? Must we not recognise *fall* to lower levels as well as *rise* to higher levels?

The question is one of fact. My belief is that this reversal of order, this downward passage in state or in status is a feature of the world in which we live, seen alike in disintegrating molecules and atoms, in degenerate organisms, in degraded minds, in debased institutions. It seems to be no less given in the evidence, as matter of fact, than is evolution. And if it be matter of fact we must realise that such it is.

Before passing on let us briefly review the position. The word "evolution" is used in two senses. To many, perhaps to most people, the more familiar sense is that in which it is said that evolution is the way in which existing animals and plants have arisen by natural descent through heredity from more primitive ancestors. If, however, we speak, as many do speak, of the evolution of molecules or crystals, there is no suggestion that they too have arisen by natural descent through heredity. Here, therefore, the word evolution is used in a less restricted sense. In that sense evolution is the building up of new wholes which are progressively higher, more complex, and richer in qualities, but which are no less new. Thus evolution is the keynote of all upward and onward advance throughout nature. For some of us it is, from first to last, subject to the directive presence of God.

I believe that this is so. But whether it is so or not, what do we find? Let me put it, somewhat picturesquely, thus: Evolution is the progressive coming into existence or being of higher and richer modes of fellowship, in which the constituent members play their parts. Of modes of fellowship there is an ascending series. In the atom, protons and electrons play their parts in one mode of fellowship. In the molecule, atoms play their parts in a new and higher mode of fellowship. In the living cell, atoms, molecules, and "colloidal units" play their parts in a far higher mode of fellowship. In our own bodies, myriads of living cells play their parts in tissues and organs which play their parts in the fellowship of the body as a whole. We thus reach the "kingdom of life." And here the reproductive cells play their parts in continuing an unbroken line of life-fellowship. Here, therefore, the doctrine of descent through hereditary transmission comes into the picture.

But there is not only progressive evolution of modes of fellowship higher and higher and yet higher. There is also dissolution of fellowship. The wholes that have been built up in evolution break down in dissolution. Some day our bodies, with their organs and tissues and cells, will break down into widely distributed molecules and atoms in sundry chemical fellowships. The life-fellowship in our bodies will no longer be the fellowship of life. This is an example of dissolution. But in our children the life-fellowship of tissues and organs continues unbroken to bear onward the torch of progress in evolution.

I have sought to show that there is abundant evidence, in the world as we know it, of dissolution—may I now say dissolution of fellowship? Without it perhaps the evolution of new and higher modes of fellowship would not be such as we find it to be. But as a matter of fact, in what we may speak of as the age-long process of the building up and breaking down of modes of fellowship, evolution has prevailed over dissolution. Were this not so the higher modes of fellowship would have passed away and would no longer exist. Were this not so we should not be here to discuss this difficult problem, or, through mental and spiritual fellowship, to contribute in some measure to progress in evolution. The world as it now is affords irrefutable evidence that evolution has prevailed over dissolution.

None the less we should realise that there is in our world, at all levels of natural events, evidence of dissolution of fellowship. Falls to lower status there are; but rise to higher status has won through. Our theme is the prevalence of evolution. And here the passage is upward to something higher. And within this progressive advance we ourselves have been caught up as active and open-eyed participators. That is where mind in evolution comes at last into the picture. Others will deal with the evolution of material things and of those no less material structures we call living bodies. The physiologist deals with the evolution of the brain. I am called on to deal with mind in evolution.

We must all admit that the body and the mind are, in millions of living beings here on earth, in some way very closely connected. We may tell a story of the body. Can one state in a few words what is distinctive of the mental story as contrasted with the bodily story? Let me try to do so. In the mental story there is *enjoyment.* You and I know what this feels like; and that is the only way we can know it. It is in common parlance pleasurable; but since discomfort and pain, though opposites, are the same *kind* of feeling, let us include these as "negative enjoyment."

In the mental story there is also that which may be called *reference.* When I see in the sky a halo round the moon and I think this portends bad weather there is in my mind reference to the rain I am led to expect. If I have to go out to a long meeting I shall take my umbrella. In that case there is *guidance* of action with reference to a possible wet night in order that I may prevent, so far as I can, the discomfort of a drenching.

Now other words than those employed here may be used. But on these terms whenever there is pleasurable enjoyment or its negative discomfort; whenever there is objective reference, as it is called; wherever there is guidance of behaviour or of conduct to the end of gaining pleasurable enjoyment or of avoiding discomfort we have characters distinctive of mind. Consider whether this brief statement, so far as it goes, is accordant with the facts of experience, and if so whether anything but a mind has these distinctive characters.

Of course all three may go together. When one is angry there is reference to some one with whom he is angry and to something he has done; one's action toward him is appropriately guided, and there is emotional feeling—pleasure or pain—which tingles in one's mind and is part of the mental story.

But though all three may go together we may distinguish them, just as we may distinguish (under reference) the colour, the scent, and the shape of a rose, though they too go together. Under this distinction mental evolution is threefold. In the ascending order of organisms (each with body-mind) from some lowly animal to man, there is, as we infer, evolution of enjoyment from lower to higher forms—from the pleasures of sense to aesthetic, intellectual, and moral joy; there is evolution of objective reference which leads up from bare sensory "acquaintance" to all that falls under "knowledge"; there is, in due course, evolution of guidance, which, by progressive steps, enables us to thread our way sure-footedly in a difficult world. All three conspire, in accordance with their level of evolution, to give the status of this or that mind in any given organism, from the lowest to the highest—conspire, too, to give the status of the mind of the individual at successive stages of its life history, say in early and later infancy, early and later childhood, adolescence, and maturity.

The threefold evolution, distinguishable under the headings of enjoyment, objective reference, and guidance of action, is threefold only in so far as we regard evolution in mind from these three standpoints. For your mind and my mind is "all three in one" and exemplifies one evolutionary advance. So, too, body and mind are distinguishable; but here and now they are nowise separable. What I have spoken of as two stories, a mental story and a bodily story, are therefore two stories, two versions of that which is given as the one and indivisible progress in the life and mind of this or that individual being.

Since, then, these two stories of one organism are closely connected, we must try to keep both in view; and since we want to get some clue to the "origin of mental species" we naturally ask whether we are justified in supposing that wherever there is life there also is mind, though perhaps in some very primitive form. Plants here present a difficulty; so let us restrict our attention to animals. Then we may say that, so far as we believe—and most of us do believe—even so simple an animal as an amoeba has something, however rudimentary, of the nature of enjoyment, and something, however incipient, of the nature of reference to its environment. Thus far we do suppose that where there is life there also is mind, though it may be a very simple form of mind. It is, then, for the physiologist to tell *his* story in terms of action and reaction under physical influence, and for the psychologist to tell *his* story in terms of enjoyment and reference.

It will, however, be noticed that what is thus attributed to the amoeba is enjoyment with reference to its environment. Nothing was said as to guidance of behaviour on the part of the amoeba. Why was this? Because opinions differ. There is divergence of view. Some of those who have studied with due care such lowly animals tell us they find quite convincing evidence that the behaviour of these organisms shows guidance of action on their part. Others say that, on the evidence as they read it, they are not prepared to attribute to a good many of the lower animals any guidance of action on their part. This raises a technical and difficult question. We cannot discuss it here; but it is pretty easy to see that if some of the lower animals give no evidence of guidance of action and a great many of the higher animals give abundant evidence of such guidance, there must be some stage of evolutionary advance at which an important feature of mind, hitherto absent, is no longer absent but very much in evidence.

What, then, is the kind of evidence? And what does guidance imply? It is guidance of behaviour and of conduct. And it seems that there are two evolutionary stages of guidance: (1) the higher, reflective or thoughtful guidance, which comes in us at the age of 2½ or 3 years and then progressively increases from about that age onward; and (2) the lower, unreflective guidance, of which we find evidence in the infant and in many of the lower animals. It is difficult to see how there could be guidance, even at the earlier stage, if there were no reference to an objective world, for it is with reference to that world that behaviour is subject to guidance. But it is, I think, easy to see that reference only to what is going on *now* would not afford what seems to be essential to guidance. It seems essential that there should be *prospective* guidance, anticipating, if only by a little, that which *will* come in the course of some established routine. How else can what *may* come be hastened or avoided by acting in this way or that? Does not all guidance imply some measure of reference to future events rendered present in expectancy, however shortsighted?

I take it that all may agree that under mental evolution we have to ascertain the manner in which conscious guidance advances from lower to higher stages, no less than the manner in which enjoyment, with reference to surrounding objects and eventually persons, likewise advances from lower to higher status. But just how guidance arises, and what is its psychological accompaniment, is perhaps the crucial question in the whole wide range of evolutionary advancement. Consider what it means. It means nothing less than the dawn of that freedom of choice which we cherish above all things. It is the very turning point in the evolutionary history of events. In that history it is of all events the greatest in promise. In human life it marks us as what we verily are—makers of a new, and, as we hope, a better world. For human guidance is always toward something more or less clearly envisaged as not yet in being, but still to be brought into being through striving and endeavour. It comes with that higher enjoyment we call joy; it comes with reflective reference. But it comes with that touch of genuine newness which characterizes every step in evolutionary advance.

In social and personal progress guidance becomes more and more and more the expression of human purpose. It is guidance in the light of deliberate and thoughtful reference, with widening range of outlook. It is guidance toward personal joy in right conduct. More than that; it is guidance toward the sympathetic rejoicing in the joy of others which characterizes love and good will. Above all it is guidance in so acting as to promote evolution and to combat dissolution. For regress there is. Our aim should be to fight it in all its forms. Here we have mind at its highest and best in social life.

In close connection with the discussion of evolution a question arises which many of us deem gravely important: What is the bearing of evolution on the religious convictions of the majority of people? Here evolution is generally taken in the unrestricted sense we have accepted. And here we find a noteworthy change of attitude. Three or four decades ago it was widely held that belief in evolution is incompatible with belief in God. One must choose, it was said, between one and the other. Much modern thought no longer regards this supposed alternative as logically sound. Some of us find no inconsistency in believing in both. Nay, more; many thinkers to-day are convinced that only in the light shed by the concept of evolution does the full richness of Divine Purpose, as thus manifested, appeal to some at least of those in whom a spiritual attitude toward God has itself been evolved.

Consider the matter a little more closely. Herbert Spencer, in 1858, contrasted "creation by evolution" with "creation by manufacture"; and even then he expressed the opinion that creation by manufacture is a much lower concept than creation by evolution. It may be said, however, that neither evolution nor manufacture express what we mean by creation. Creation, or as it used to be called, "special creation," means, it will be said, sudden bringing into being by unconditional fiat. As a typical example of creative fiat take, "And God said, Let there be light, and there was light." Extend this: Let there be things; let there be plants and animals after their kinds; let there be man. Such was an early expression of creative fiat. It was poetical in the fine sense that

> God on His throne is eldest of poets;
> Unto His measures moveth the whole.

Noteworthy is that wonderful touch of spiritual insight, fitly given first place in the Hebrew Scriptures: Many instances of creative fiat, but One God whose Purpose is thus manifested.

Turn now to modern thought. It is still open to us to couch ultimate explanation in like terms: Let there be electrons; let there be atoms; let there be molecules; let there be crystals; let there be life; let there be man with knowledge of good and evil. Each in turn is severally an expression of that which, for lack of a better phrase, we speak of as Divine Purpose, freed from the temporal limitations of human purpose; but each is an independent and unconditional expression thereof.

It may still be asked, however, whether it is not open to us to interpret *also* in the light of all that science has taught us—that is, in terms of interdependence and the linkage of the natural events themselves—to trace the steps through which and the conditions under which this or that item in the list from atom to man came gradually into being. Is it not open to us to accept evolution without rejecting Divine Purpose? The concept of fiat—if it is still to be retained as helpful—then takes the form: *Let there be one natural plan of evolutionary progress exemplified throughout in many and diverse ways.* It is because I have been led, through my survey of what seem to be the patent facts, to find one evolutionary plan as the manifestation of one Divine Purpose (difficult as this may be to define) that I prefer the unrestricted usage of the word "evolution."

On grounds such as these it may be urged that acceptance of unrestricted evolution though many ascending grades, reaching its culmination in the highly-developed mind that plays so great and increasing a part in later evolutionary progress, is not incompatible with belief in God as manifested in all advance from lower to higher.

In any case many evolutionists hold this belief with sincere and enhanced conviction. What, for one of them, does this imply? It implies *reference* to God as object of spiritual contemplation; it implies *guidance* of conduct in the light of this reference; it implies *joy* in attaining such ends as are deemed to be consonant with Divine Purpose. But this joy, this guidance, this reference, are themselves here and now in process of evolution. In daily life they sustain worthy endeavour. They may rise from lower to higher.

We speak of this kind of advance in human life as illustrating nearer approach, with joy under guidance, to truth, beauty, goodness, and above all, to love and good will. Here mind in evolution reaches its highest expression. And even if it be part of our belief that we are thus in touch with the eternal verities, it may also be part of our belief that closer approach to their realisation in human affairs is seen in that progressive process in time which we call evolution.

C. LLOYD MORGAN

CREATION IS INVOLVED IN WHAT GOD HAS EVOLVED

"The world has been evolved, not (specially) created; it has arisen, little by little, from a small beginning at an almighty word. What a sublime idea of the Infinite might of the great Architect, the Cause of all causes, the Father of all fathers, the Ens Entium! For if we would compare the Infinite it would surely require a greater Infinite to cause the causes of effects than to produce the effects themselves."—Erasmus Darwin (Grandfather of Charles Darwin).

AND THEN RETURN

"It is absolutely certain that we are in the presence of an Infinite Eternal Energy from which all things proceed."—Herbert Spencer.

CUMULATIVE EVIDENCE FOR EVOLUTION

No greater mistake about evolution could well be made than to limit its application to living organisms. There has undoubtedly been as real an evolution of the Cosmos, of the solar systems, of the earth and other planets, of the molecules, and of the atoms as there has been of organisms. All of these have much in common. In none of them is there any fixity or stability; in all of them there is rhythmic and orderly change. In none of them does the course of change proceed steadily in one direction. On the contrary, it commonly seems to proceed from states of less complexity to states of greater complexity and then to revert to states of less complexity. For example, according to the latest theory, a sun such as our own—and there are hundreds of millions of these in our own galaxy alone—is believed to have had many vicissitudes during its lifetime of quintillions of years. In the course of its wanderings through space it may come relatively close to another passing sun and during this passage give birth to a family of planets, each of which is its child. As one would expect of a sun's child, each planet has a long period of growth, a period that lasts for billions of years. Sooner or later, however, our sun may come again within the gravitational reach of another sun, the calculated average interval between such approaches being, in round numbers, a billion times a billion years. When this happens, the first family of planets will be broken up and another one born. This rhythm may go on forever, so far as we can tell, for there appear to be no agencies tending to put an end to it.

Each star or sun, apart from these incidents associated with the origin and evolution of planets, has a long, slow evolution of its own. Each is at first a young sun, very tenuous, relatively cool, and blue-white in color. As it grows older it becomes hotter, denser, and yellowish in color. With increasing age it becomes progressively denser and cooler and changes in color from lighter to darker red. With this progressive increase in density the constitution of its atmosphere undergoes remarkable changes. In the young suns the atmosphere consists of numerous lighter elements and compounds that can exist at relatively low temperatures; in the somewhat older suns, which are intensely hot, the atmosphere contains only the lightest and simplest atoms, such as hydrogen and helium; in the old, red suns the atmosphere includes not only atoms of the heavier elements, but various compounds. The density of some of these aged suns is amazing. Astrophysicists estimate that a cupful of material from one of these old red suns, if weighed on the surface of our earth, would scale twenty-five tons.

A sun endures so long and changes so slowly that an ephemeral being like a man can observe no change in it. We know, however, that our own galaxy is made up of suns of all grades of brightness and density, and we therefore infer that suns have a regular course of existence—an evolution. Our own particular sun is middle-aged, verging on senility. To an observer living on a planet in another solar system it would appear as a reddish-yellow star, relatively a dwarf as compared with many of the giant suns in our galaxy.

A sun is at all times giving off enormous amounts of energy. This activity alone would cause it to change progressively and continuously. All the evidence in our possession indicates that, apart from occasional relatively sudden, more or less accidental short cuts, the course of solar evolution is almost inconceivably slow.

Passing from suns to lesser material units, let us consider the evolution of planets. The lifetime of a planet as compared with that of a sun is short, and its evolution is correspondingly much more rapid, so rapid that many of the events of its career can be noted by man. In our own planet, for example, we have observed changes in the levels of continents, changes due to earthquakes, floods, and the viscissitudes of climate. Similar but far more extensive changes are recorded accurately in the strata of the earth's crust.

An intelligent perusal of the rocky pages of our earth's historical record shows that the hills are not eternal but are periodically coming into being and passing away; that the oceans change their depths and contours; that the continents join hands for a time and then part company; that parts of continents become islands and that islands become attached to continents. If viewed by a being whose time passes as slowly as it passes for the sun, the earth would appear to be in a continual state of flux. It would seem to pulsate like a gigantic heart as the continental and oceanic areas periodically expand and contract. During the periods of relatively expanded continents and contracted seas the lands in many regions are high and mountainous. These high lands and mountains undergo a long, slow period of gradation, during which much of their solid material is washed into the sea. The filling up of the sea helps to make it overflow upon the lowest parts of the continents, and thus the area of the oceans increases and that of the continents decreases. Before this goes very far the increased weight of the sea floor and the lessened weight of the continents brings about another squeezing up of the continental blocks and a sinking of the sea-levels, and a new pulse beat of the earth takes place.

When the continents are extensive and high the climates are zonal—that is, there are zones of different climates in different regions. The climate is cold at the poles, warm at the equator, arid in central regions, and moist along the coasts. When the continents are contracted and worn down low the climate is non-zonal and equable, the conditions of life are easy, and evolution is relatively slow. The rocky strata of the earth's crust show clearly that there have been several great continental pulse beats. A section of the walls of the Grand Canyon, for example, affords an opportunity to read the book of geology at a place where the earth has herself turned back the pages. To one who has learned to read the language in which it is written, nothing could be more certain than the story thus revealed. It is a true story, unspoiled and unedited by the hand of man. Those who have become experts in reading the record of the rocks agree in the interpretation of its pages, at least as to its main plot. About some minor details different versions are offered. It is clear, however, that since the earth reached its present diameter, at least a billion years ago, there have been no less than half a dozen major pulse beats of the earth; and numerous minor rhythmic changes have been superimposed upon the major rhythm. It is also certain, as shown by fossils, that the earth has been the abode of life during numerous physical and climatic changes of far-reaching extent and of great severity.

Leaving now the exceedingly large units of the universe, let us put on our shrinking caps and magically pass to the exceedingly small world of the atom. Modern physicists have revealed to us that the atom, once thought to be the smallest thing in existence, is really much like a miniature solar system, composed of a central relatively massive body around which revolve one or more planetary bodies, each occupying perhaps only about a millionth part of the space occupied by the smallest atom. An atom of hydrogen is the simplest atomic system known. It is composed of but one proton, or positively charged central particle, and but one electron, or negatively charged particle, which revolves in an orbit about it. The speed of the revolution of the electron is so tremendously great that it is practically everywhere at once within its orbit. The distance between the proton and the electron seems so enormous as compared with the minute size of these particles that one is forced to the conclusion that the inside of the atom is mainly empty space. One may get a more concrete idea of the relative sizes and distances within the atom by comparing the atom of hydrogen with the earth and the sun. It may be said, speaking broadly, that if the orbit of the electron about the proton in an atom of hydrogen were enlarged to the size of the orbit of the earth about the sun the electron would have a diameter about equal to that of the sun and the proton a diameter about equal to that of the earth. This somewhat topsy-turvy relation is due to the fact that the proton, though ever so much smaller than the electron, is nearly two thousand times as massive, or heavy.

Other atoms are far more complex than the hydrogen atom, some of them containing over two hundred times as many protons and electrons. No matter how large and complex an atom becomes, it includes no other kinds of particles than those contained in the simplest atom. All differences in the properties of the elements are due to the number of and the variations in the arrangement and configuration of these ultimate particles. The nucleus of any other atom than that of hydrogen is composed of both protons and electrons, firmly organized into a relatively stable core, around which revolve a few or many planetary electrons in one, two or several shells, each capable of housing a definite number. Only the electrons in the outermost shell determine the chemical characteristics of the various atoms of the elements.

Some of the most complex atoms, such as those of uranium, thorium, and radium, are radioactive; that is, there is a sort of unrest in the nucleus which operates to break down the equilibrium existing among the protons and electrons and results in the shooting off, at tremendous velocity, of electrons and of groups of protons and electrons from the nucleus. The so-called Alpha rays given off by radioactive substances are composed of particles identical with the stripped nuclei of the helium atom, one of the lightest and simplest atoms. The emission of these rays is nothing more or less than a process of evolution of elements, one element becoming transformed into another. By radioactive disintegration the most complex elements, such as uranium, thorium, and radium, are reduced slowly and by distinct steps to elements of less complexity and greater stability. Thus when an atom of radium loses one Alpha particle it is reduced to radon, an extremely inert gaseous element. Radon goes over by another step into polonium and lead, and radioactive lead, by a further change involving the capture of another electron, becomes the element bismuth.

Physicists are just at the beginning of their program of transforming the elements or atoms into one another, and doubtless their future research will disclose many startling transformations. The evidence so far available points to the conclusion that what we once considered the most fixed, the most immutable units of the universe are far from stable—they are undergoing orderly transmutation from more complex to simpler forms. The transmutation seems very slow to us, but expressed in terms of cosmic time it is really rapid.

The question naturally arises whether this process of atomic degradation is a one-way process, destined in the end to reduce all matter to its simplest form. The most recent discoveries of physicists seem to answer this question in the negative. There is a well-defined belief among experts in these matters that processes the reverse of those described above, involving a synthesis of simpler forms of matter into more complex, are going on out in the remotest interstellar spaces. Professor Millikan has detected vibrations emanating from these outer spaces—vibrations whose frequency is not even approached by any known earthly or solar phenomena—that are interpreted as evidences of immensely energetic synthetic processes involving the return of matter from its state of ultimate disintegration to conditions of integration and complexity. These observations seem to indicate that atomic evolution, like other phases of evolution, is rhythmic and orderly. Such a rhythm would seem to have no beginning and no ending. Perhaps our ideas of beginnings and endings are due merely to the limited functioning of our human brain mechanism.

In the hands of the expert the spectroscope is seemingly a magical instrument. By its help he can reach out and measure the distances and the diameters of the remotest stars, and even of the outer galaxies; he can use it as a long-range thermometer with which he can read the temperatures of the most distant suns; with it as a speedometer he can calculate the velocity of any of the heavenly bodies; and by its aid he can determine the chemical composition of the external parts of the suns almost as accurately as if he had them in his laboratory.

Spectroscopic analysis indicates that there is an orderly and systematic progression in the temperature and composition of the suns from the giants, or young suns, to the dwarfs, or old suns. Along with these evolutionary changes in the general character of the individual sun there is a parallel evolution of the elements. Thus it appears that evolution in the most minute units is definitely linked with evolution in the largest units. One evolution is obviously causally related to the other.

This conclusion leads me to venture upon the bolder statement that all evolution is in the end one vast universal coördinated process. We may be able to view one or two of its various aspects as though they were set apart from the rest, but any adequate study of one phase of evolution sooner or later leads to the conviction that it cannot be understood as a self-contained process or mechanism but can be made intelligible only by considering its relation to other processes or mechanisms. In the end we are inevitably driven to the conclusion that all nature is an organized system and that whatever happens in one realm is related to all other realms, and thus to the whole. Such a view as this leaves one with a feeling of awe in the presence of that vast unity we call Nature. There is room here, if anywhere, for a scientific concept of Deity, a central immanent power back of all these coördinated activities, from the smallest to the greatest.

If, as the astronomers, physicists, and chemists tell us, all lifeless nature is engaged in a ceaseless swing of intense activities; if the sun is growing older and changing its character with every succeeding day; if the earth has had its ups and downs, with numerous radical changes of climate; if, as the rocks tell us, life originated prior to a long series of these great environmental upheavals, it is indeed difficult to believe that living organisms, the most plastic of all natural units, should have remained fixed amidst the vast flux of world changes. In fact, there is between the records of geologic change and those of biologic change the closest parallelism. Whenever the rocks tell us a story of relatively sudden continental uplift, with its associated climatic changes and stormy times, we find a corresponding adaptive change in the organisms preserved in the rocks. Whether the organisms respond directly to the changes in environment, or whether there is merely a change in standards of survival and an elimination of the less adaptable forms, we do not certainly know. Nothing could be clearer, however, than that there is a causal relation between organic and inorganic rhythms. Thus once more the unity of the whole process is impressed upon our minds. Organic evolution must be viewed not as an isolated process, but as an integral part of a vast system of orderly change.

No one line of evidence of organic evolution can possibly be conclusive in itself, though to the expert in each field his own data seem to need no outside support. The real proof of the validity of the concept of evolution lies in the fact that all lines of evidence point in exactly the same direction and are fully consistent with and corroborative of one another. Not only is this so, but each kind of evidence throws light upon all the other kinds. The obscure spots in one field are illuminated by facts derived from other fields. Thus the proof of organic evolution is *cumulative.* Before Charles Darwin recorded the results of his epoch-making work the known facts supporting the theory of evolution were few and unorganized. It is to the everlasting credit of Darwin that he amassed and organized so large and conclusive a body of evidence of evolution that he practically established the validity of the theory single handed.

The real test of the value of a theory, however, is not that it explains and rationalizes only the particular body of data it was devised to explain. It would be a poor theory that did not agree with the facts upon which it was based. A good theory must meet and explain all new facts within the scope of its applicability that come to light subsequent to its proposal. Since Darwin's *Origin of Species* was published ten times as many new facts about organic nature have been discovered as were known to Darwin. One after another new and radical discoveries that involve profound alterations in our fundamental views of life have been made, but none of them have in any way shaken our confidence in the theory of evolution. Had any new discovery in the least weakened or run counter to that theory its ever-watchful enemies would have pounced upon the discovery and exploited it to the fullest extent. But no discovery since Darwin's time has done other than strengthen and confirm the theory. The more specialized a branch of biology becomes, the more useful and necessary is the concept of evolution. Those that are most expert in the advanced and technical branches of biology are the most ardent advocates of evolution, for their data demand an evolutionary interpretation.

The evidences of evolution have been piled up year after year until their sheer mass now overwhelms all intelligent opposition. One who exposes himself open-mindedly to the evidences can no longer believe in a static world inhabited by fixed and unchanging species.

If evolution were a false doctrine, it should be easy to refute it by bringing forward facts that would contradict it. From time to time facts or alleged facts supposed to conflict with the principle of evolution have been brought forward by those who sought to overthrow it. Invariably, however, a more exhaustive study by trained scientists has shown that not only are the facts cited not contrary to evolution, but that they tend strongly to confirm it. To-day no adequately studied facts are out of accord with the theory of evolution, and thousands upon thousands proclaim its truth. What more cogent proof of a theory can one ask?

An excellent example of the way in which unexpected discoveries in a new field have supported and confirmed the theory of evolution is seen in the new science of serology, or blood tests. Before anything was known about the specific chemical constitution of the blood, animals had been classified into phyla, classes, orders, families, genera, and species. The method used was the method of homology—that is, groups of animals that were most nearly similar in structural pattern, not only in the adult form but throughout the course of embryonic development, were believed to be most closely related and were placed in the same species. Animals differing in details but having the same general features were placed in the same genus, family, order, class, phylum, according to the degrees of their structural resemblance. A comprehensive system of classification has thus been built up that is believed to constitute a sort of pedigree, or ancestral tree, of animal life.

A decade or so ago an entirely new and highly refined method, quite unrelated to the method of homology, was devised for testing animal relationships. This is the so-called blood precipitation method, which depends upon the fact that the blood of an animal is a sort of quintessence of its chemical composition. Thus the blood of all human beings has a highly specific chemical constitution differing from that of all other species. The same is true of the blood of the dog, the horse, or any other animal. The degree of chemical resemblance and difference in the blood of different animals may be measured quantitatively with the greatest accuracy. Assuming, then, that the degree of chemical resemblance in the blood of different animals will indicate the closeness of their genetic relationship, we should be able to classify the whole animal kingdom by means of resemblance in blood.

The technique involved may be elucidated by a single example. If we wish to find out the degree of kinship of man to any kind of lower animal, we proceed as follows: From a quantity of human blood we draw off only the clear, colorless serum. This we inject at intervals into the veins of a rabbit, for example. In time the rabbit's blood becomes charged with a specific antibody for human blood and the serum from the blood is known as anti-human serum. This rabbit serum thus produced can now be used as a chemical reagent for testing the affinities of any other blood to human blood. If a few drops of it are placed in a test tube full of human serum a heavy white precipitate is immediately formed. If placed in the serum of a gorilla or of a chimpanzee a definite precipitate is formed, but it is less abundant and it forms more slowly than if human serum is used. No other species tried gives so positive a reaction with anti-human serum; but the baboons, the New World monkeys, the marmosets, and the lemurs (all Primates) react less and less readily in the order mentioned. All Primates show a stronger affinity than any other mammals for human blood, but if larger amounts of the reagent are used and more time is allowed for the reaction, the degree of affinity between man and all other mammals may be shown. Moreover, the order of closeness of relationship corresponds to that already worked out by the method of homology.

Most of the larger groups of animals have been investigated by this method and the most significant relationships determined are the following:

1. The birds show close relationship to the reptiles.

2. The king crab, *Limulus,* is more closely related to scorpions and spiders than to any true crabs.

3. The whales, whose affinities among mammals have long been a problem, show an unmistakable affinity to the hoofed mammals, especially to the swine.

4. All Primates show closer affinity to one another than to any other mammals.

5. Similarly, all Carnivora are more like one another in blood than they are like other mammals.

When the system of classification based upon blood tests is compared with that based upon homologies it is found that *the two corroborate each other in all essential respects.* Where the method of homology had left the affinities of an animal somewhat doubtful, the blood test has been a valuable check upon earlier findings. Some relationships that were only doubtfully hinted at by homology have been definitely confirmed by blood tests.

One of the most conclusive evidences of the essential truth of the concept of evolution is that two utterly different methods of testing the relationships of animals should thus be in close agreement. If the blood-test method had given a different set of relationships than those indicated by homology we should doubtless have lost confidence in the validity of one or both methods, and our confidence in the principle of evolution would be to some extent shaken. To the same degree, then, that a disagreement in the results of the two methods would have weakened our confidence in evolution should not the close agreement of the two methods strengthen our confidence in it?

Another example of the way in which the evidences from diverse fields of science converge upon one conclusion, and only one, is to be found in the relationship of birds and reptiles. Studies of the comparative anatomy of adult birds long ago led to the belief that birds and reptiles have more in common than any other two classes of vertebrates. A study of comparative anatomy alone shows that birds are, as has been said, little more than "glorified reptiles," differing from other reptiles mainly in the possession of feathers and wings. Blood precipitation tests support this conclusion; the blood chemistry of the two classes indicates that they are closely related. Embryology reveals the fact that birds, long before they are hatched, have the beginnings of typically reptilian teeth, which never reach maturity. The eggs, embryonic membranes, and indeed the whole course of the embryonic history of birds, is strikingly reptilian. In fact, it is only in later stages of embryonic development that the true avian characters begin to appear. The most characteristically avian feature of a bird consists of its feathers; but even these show by their development that they are no more than finely subdivided reptilian scales.

If, then, birds are specialized descendants of reptiles, it is obvious that there must have been transitional stages leading from reptiles to birds; and it is just here that palaeontology furnishes the evidence that settles the question. In a deposit of Bavarian shale there have been found two nearly complete and well-preserved fossil specimens of a kind of animal that is hard to classify as either reptile or bird, for it obviously possesses some of the features of both. This extinct animal, which is known as *Archaeopteryx,* is an animal halfway between a reptile and a bird. It had true feathers and it had fore limbs that are half wings and half fore legs, each having three long, prehensile fingers. It had a long, slender lizard-like tail, on which there was a lateral fringe of large feathers. The head was essentially reptilian, having no horny beak but a full set of reptilian teeth. What better evidence could one wish that the birds have been derived from reptilian ancestors? Is it even remotely probable that the story of evolution told independently by all these natural records is false? Nature does not lie. Here, as elsewhere, several of nature's witnesses tell the same consistent, straight story, presenting cumulative evidence that cannot be refuted.

Biology furnishes many examples of the convergence of evidence upon a common conclusion. Only a sample or two have been offered to illustrate the fact that the proof of evolution, though somewhat indirect, is conclusive. It is the sheer mass of converging and cumulative evidence that in the end wins the day for evolution.

Though most of the evidences may be called indirect, this word cannot be applied to the evidence derived from genetics in the study of evolution going on to-day. The modern geneticist breeds under observation and control huge populations of rapidly breeding animals and plants. Under his very eyes there come into being scores and hundreds of new or changed types of individuals that pass their peculiarities on to their progeny according to definite laws of inheritance. Most of the changed types (mutants) are inferior in various ways to the typical individuals of the species. Some mutants are so weak or defective that they die young and leave no offspring. Occasionally, however, a mutant appears that possesses a new character or set of characters that constitutes an improvement. Such a new character persists and becomes incorporated in the hereditary complex of the species. Evolution can thus be seen to proceed by the production of large and small hereditary variations and the persistence through heredity of the good or relatively harmless mutations.

If a species becomes geographically separated into two groups, these groups tend gradually to diverge more and more. There are at least two reasons for this divergence: first, a group of individuals that becomes isolated from the main body of its species will tend to form a local race characterized by racial differences; second, if the environments of two groups thus separated are different, the standard of survival will be different, not only for the older characters but for the ever-recurring mutations as well. In the course of time these divergent forces inevitably create distinct species.

In his studies of contemporary life the geneticist has actually observed the process of evolution in operation. The processes observable to-day, if projected into the past, would be adequate to account for the evolution that has taken place in the past. Science had taught us that the present and the past are one; if we can analyze the present we have the key to the past and to the future. Evolution is obviously going on to-day. What better proof than this do we need for our belief that evolution has gone on in the past?

In conclusion let us say that the principle of evolution is so well established by the amassed evidence derived from every field of science that it has come to be regarded in scientific circles as one of the great laws of nature, ranking with the law of gravitation in scope and validity. And now a word for the theologian: Evolution no more takes God out of the universe than does gravitation. Both these great principles are mere manifestations of the grand strategy of Nature. They indicate the methods used by the ruling power back of the universe. The theory of evolution, as has often been said, does not deny creation; it merely explains the method of creation.

HORATIO HACKETT NEWMAN

TODAY I LEARNED THE EGG WAS LAST

Which was born first the chicken or the egg. If you understand evolution it's really pretty easy to answer this question. I might feel at first sparks only created sparks, and then some cooled down and started hatchlings, only a glowing creation, and then less and less until they actually had a youngster they could touch; not hot at all. Just think pretty quick they were having babies out of their mouths and all sort of things. Sometime later a feeling of getting rid of the burden's bungle-some weight would make raising young alot easier and besides that the mother can get the old man to keep the egg hot, that's if we'll call them eggs, warm while she goes shopping. So it only took another million years to get it, the yolk and all. So there you have it; remember this thought, in fact I think they were the ones that God put out eating dust for the rest of his life for giving Eve the apple she gave to Adam.

BILL HUTH

WE ARE!

"The purpose of science is to develop, without prejudice or preconception of any kind, a knowledge of the facts, the laws and the processes of nature. The even more important task of religion, on the other hand, is to develop the consciences, the ideals and the aspirations of mankind. Each of these two activities represents a deep and vital function of the soul of man, and both are necessary for the life, the progress and the happiness of the human race.

"It is a sublime conception of God which is furnished by science, and one wholly consonant with the highest ideals of religion, when it represents Him as revealing Himself through countless ages in the development of the earth as an abode for man and in the age-long inbreathing of life into its constituent matter, culminating in man with his spiritual nature and all his Godlike powers."—Dr. Robert A. Milliken, in *Science*.

MAKE ROOM FOR WORDS

Recently, in this book, I tried to bring the reader's attention to the tongue. While reading THE EVOLUTION OF THE BRAIN. I was paying close attention to the comparison of the lion, gorilla, and the human brain and the various sizes of the brain of each. "The difference in size is confined almost wholly to three areas," G. Elliot Smith states, and continues, "and it is profoundly significant that the areas which reveal the enormous expansion in the human brain are precisely those that attain their maturity latest in the human child." The brain of the large ape and the brain of the human child are nearly the same size. The areas of touch, visual, and acoustic are just as well developed in the ape as in man. However the latest areas to mature are enormously bigger in the human brain. The only contrast between the human and the simian (ape) brain is that certain areas in the brain of man are enormously bigger than the corresponding areas in the brain of the ape. The structure of these corresponding parts - perhaps the most amazingly complex pieces of living machinery - is so similar that even the most experienced anatomist is unable to distinguish between them.

I feel that the reason for these certain areas to mature last were the result of the child beginning to use his tongue as a method of speech and I am also positive the storage capacity for the oncoming vocabulary could very well be the result of the larger growth in the human brain.

BILL HUTH

TRUTH FROM THE HEART
ALL CREATION SPAWNS FACTIONALISM

This book contains ninety-seven short, shorter and one line articles. These works were written as concise; and to the point as possible by each author. Every effort has been made by each of these fifty-three voices, who contributed to this book, to interpret, clarify and illuminate the subjects of each article as best they could, making every effort possible so the reader might fully understand it's substance.

About the voices who each contributed his time and effort to make these works come to life. In the background of each individual there are many, many references. Perhaps the earliest could be the works of John Ray, born November 29, 1627. He wrote "WISDOM OF GOD MANIFESTED IN THE WORKS OF THE CREATION" *in the year 1691 A.D.; William Paley, who was born in 1743, used Ray's manuscript for reference to write* "NATURAL THEOLOGY". *Paley expounded that form of the teleological argument for the existence of God comparable to the argument that the existence of a watch implies the existence of a watchmaker. This book was one of the great influences on the life writings of Charles Darwin. Undoubtedly Darwin's writings, together with many unmentioned voices of the past, were referred to in making up these ninety-seven articles. A few of our authors were born in the seventeen hundreds. Most of them in the eighteen hundreds, a couple or so in the nineteen hundreds, and of course King Solomon, who deserved the cover: born approximately one thousand years B.C.*

THE END CAN ONLY MEAN THE LAST PAGE:
CREATION IS FOREVER!

CONTRIBUTORS

Agassiz, Louis
Bather, Francis Arthur
Berry, Edward Wilbur
Boyesen, Hjalmar Hjorth
Buffon
Burroughs, John
Conklin, Edwin Grant
Darrow, F.L.
Darwin, Charles
Darwin, Erasmus
deBorlodot, H.
Dendy, Arthur
Dorsey
Gager, C. Stuart
Goethe
Gregory, John Walter
Gregory, William King
Holmes, Samuel Jackson
Huth, Bill
Huxley, Julian Sorell
Jennings, Henry Spencer
Jordan, David Starr
Loomis, Frederick Brewster
Lubbock, Sir John
Lull, Richard Swan
Lyle

CONTRIBUTORS, continued

Maeterlinck
Mason, Francis
McBride, Ernest William
McCloud, James
Meyer, Joseph
Milliken, Robert
Morgan, C. Lloyd
Morley, Lord
Newman, Horatio Hackett
Newton, Sir Isaac
Olson, H.F.
Osborn, Henry Fairfield
Parker, George Howard
Poultin, Edward Bagnall
Rickey, Branch
Scott, William Berryman
Shakespeare
Sherington, Sir Charles Scott
Shipley, Sir Arthur Everett
Smith, G. Elliot
Solomon
Spencer, Herbert
Thomson, J. Arthur
Wallace, A.R.
Watson, David Meredith Sears
Wheeler, William Morton
Woodward, Sir Arthur Smith